Modernity of India : Ambiguities and Deformities

Modernity of India : Ambiguities and Deformities

Edited By :
Ashim Kumar Sarkar
Kaushik Chakraborty
Manas Dutta

Setu Prakashani

Modernity of India : Ambiguities and Deformities
edited by Ashim Kumar Sarkar, Kaushik Chakraborty, Manas Dutta

First Edition : September, 2014
Cover Design :Ebong Advertising Solutions pvt. Ltd.

ISBN: 978-93-80677-68-2

Published by:
Archana Das & Subrata Das
on behalf of **Setu Prakashani**
12/A Sankar Ghosh Lane, Kolkata-700006
Phone : +91 33 2219 0704, +91 94330 74548

Outlet :
2, Shyama Charan Dey Street, Kolkata - 700 073

Delhi Branch Office :
A-43, Mahendru Enclave, Galli No. 3,
G.T. Karnal Road, P.O. : Azadpur, Delhi - 110 033
e-mail : setuprakashani@gmail.com
Website : www.setuprakashani.com

Price : Three Hundred and Seventy Five

Setting by :
Cynosure, 3/37A, Bejoygarh, Kolkata - 700 032
Printed by :
Repro India Ltd.

Dedicated to the Revered Memory of
Durgakinkar Bhattacharya
Principal
Malda College (1947-1977)

Acknowledgement

We wish to take this opportunity to express our great indebtedness and heartiest of thanks to all those whose cooperation and encouragement enormously helped us to finalise this volume. Among them at first discernibly comes University Grants Commission of India (The original title of the Seminar on the basis of which the UGC sanctioned grant was 'Interrogating and Exploring Hindustan, British India and India, 1556-2000' orgainsed by the post-graduate department of History, Malda College, in collaboration with History Department of Jadavpur University between 29 and 30 April, 2011.) and Indian Council for Historical Research whose generous grant helped us to successfully conduct the National Seminar in the Post-Graduate Department of History, Malda College. This proceeding volume owes its origin in that seminar. We are also thankful to the Ministry of Information and Culture of Government of West Bengal for their Grant. We sincerely, thank the history department of Jadavpur University for their co-operation.

We are thankful to several personalities for their co-operation, contribution and service. We are thankful to Dr. Tapash Chatterjee, Former Registrar of University of North Bengal and University of Gour Banga. Thanks is due to Mr. Santosh Chakraborty, President, Malda College and Dr. Aniruddha Sengupta, former Principal, Malda College.

It is true that this volume have seen daylight because conference organizers got affectionate and sincere cooperation of Professor Ananda Gopal Ghosh, Professor Ranjan Chakrabarti,

Honourable vice-chancellor of Vidyasagar University, Professor Smriti Kumar Sarkar, Honourable vice-chancellor of University of Burdwan. Thanks is due to Professor Sajal Nag, Professor Rila Mukherjee , Professor Arabinda Samanta, Dr. Bipasha Raha, Dr. Karubaki Dutta, Mrs. Bidisha Chakraborty, Mr. Utpal Roy and all the other presenters and participants.

No word is sufficient to express our sincere thanks to Mr. Soharab Ali for his efforts, cooperation and affectionate help. By delivering the vote of thanks he helped in successfully finishing the seminar.

We gratefully admit that the conference happened successfully because we received generous support from the colleagues of Malda College. Special thanks is due for Dr. Naresh Chandra Roy, Dr. Tapan Kumar Mandal, Mr. Ashim Bagchi, Dr. Abhijit Roychowdhury. Thanks is due to Mr Debapriya Basak who smilingly undertook the difficult task of proof reading. We are also thankful to Mr. Krishnayan Dasgupta, Mr. Shirshendu Raha, Dr. Pinaki Roy, Mr. Chandan Roy, Mr. Tanmoy Pandey, Mr. Rhitobroto Goswami, Mr. Matin Ahmed, Mr. Sutanu Bhattacharya, Mr. Akhil Roy, Mrs. Sneha Sanyal, Mrs. Monalisa Laha, Mr. Hoque Sabiruddin, Mr. Arijit Mitra, Mr. Pothik Sarkar, Mr. Sahadeb Roy, Mr. Kartick Roy and to several of our colleagues and staff of Malda College.

Finally we also wish to express thanks to Mr. Subrata Das of Setu Publishers who expressed willingness to publish the work. We must say that whatever deficiencies remained in this work are due to our inability and limitation.

Ashim Kumar Sarkar
Kaushik Chakraborty
Manas Dutta

CONTENTS

Editorial

Paradoxes of Modernity

Hindustan of Mughal period has been completely changed when it became "India" under colonial rule. Mughal legacy was discontinued because the colonial state was a commercial state. Being guided by commercialism the colonial state introduced changes in every sphere. Change happened in land revenue system, in forest policy, in armed forces, in communication and so on. The indigenous society, specially peasant community on several occasion violently reacted against the changes introduced under colonial rule. That gave birth to a long legacy of anti-colonial resistance.

The fountainhead of all these changes initiated with the foundation of a "state" by the colonial power. As Peter Robb has observed that the chief colonial institution and creation was the state. [1] The most Fundamental change with far reaching effect which this state introduced was private property in land through new land revenue settlements from late 18th century onwards. Corroborating to this system came a new legal framework, administration and police etc. [2] Thus, the state originated under colonial rule became a completely different entity in comparison to the pre-colonial state. Utilitarianism, specially the ideas of James Mill established the idea that Mughal state was a version of despotism which was succeeded by a rule far more advanced than the Mughal state. [3] Subsequently by mid 19th century the idea of a development regime inaugurated under the colonial rule has percolated in the thought and worldview of Indigenous intelligentsia. [4]

In late 1840 Akhyaya Kumar Dutta, a leading intellectual figure of Bengal has observed that the pre-colonial state was an oppressive state but colonial India ended that system of oppression. [5] Similarly in late 19th century another leading intellectual Bhudeb Mukhopadhyay has observed the pre-colonial rule as despotism succeeded by a regime of freedom under colonial rule. [6] Several contemporary works, articles, speeches are replete with such ideas. This idea continued to exist and thrive in 20th century also. The paradox is that whereas nationalist freedom struggle wanted a complete end of the colonial rule but nourished the desire of capturing the power of the "state" created by the very colonial rule itself. [7] The only reason was that despite of its oppressive character the colonialism created "state of India" has been viewed and perceived as "modern".

The history of colonial India shows that on every sphere of policy implementation as well as formulation and building up of administration cum institutional framework the colonial state gave priority to commercialism or efforts to continue its existence. Colonial project of modernity bears no resemblance with the classical western European modernity which has been always promoted as a parameter of modernity. As elsewhere it is observed by modernity the colonial state promoted application of technology with a superficial urbanity [8] and thriving of commercialism. This is not what enlightenment, industrial Europe understood as modernity. [9] As A. Raghuramaraju has observed that in Western Europe the concept of modernity was consisted of autonomous individual, instrumental rationality and freedom. [10] The modern society, as Raghuramaraju further observed, has achieved a temporal ordering of traditional and then modern.[11] Raguramaraju concluded that this kind of temporal ordering does not existed in societies like India.[12]

In Western Europe modernity emerged from within the society. The rejection or modification or adaptation or disinheritance of old/pre-modern happened as a process within the respective western nations and societies. In the instance of India the whole version of modernity came from above as a western concept contrasting the pre-colonial but the concept have no indigenous root. The indigenous response to this version of modernity was not monolithic. Rather, it ranged from uncritical acceptance to passive and active resistance, revivalism and in some instances an effort to fit the modernity with indigenous tradition so that it gets an appearance of not purely western in nature. Thus, as Partha Chatterjee has observed that Indians acted as consumers of the idea of modernity produced in the west.[13] From western colonialism came an idea of modernity in a society like, India which has been perceived from the western view as pre-modern/backward as it maintained ties with old. As, Dipesh Chakraborty has observed that intellectuals of India had an ambivalent relation with the past which shaped the very idea of modernity in a peculiar way.[14] For this attachment with the past, the Indian perception of modernity, as Partha Chatterjee has observed, became radically different from the version of western modernity.[15] To sum up, in this way there emerged a modernity which does not emerged after denouncing or reshaping its pre-modern, nor was it a complete photocopy of western modernity.

From early 19th century what was visible was that there was the structure of a state resmbles with modern European nation-state. The state created administration and institutions which again resembled with modern European nation state which even facilitated participatory politics by 1930's. But at the core lay the guiding principle of colonial political economy, i.e., sub-serving everything to commercialism. So, what emerged was not modernity. The present volume is a modest attempt to throw light on this issue. The subsequent articles are varied in nature due to their issues, areas and content. Virtually none of the articles directly or even indirectly dealt with the idea of modernity or the debates associated with it. But what becomes visible from the articles is the anachronies and paradoxes of policies and ideas implemented by the colonial state which always differed from the earlier sub-continental tradition in every sphere.

The first section is consisted of articles by Anjan Chakrabarti, Sahara Ahmed and Wahengbam J. Singh under the title of *Trajectories of development Agriculture, Environment and Communication.* Anjan Chakraborti discussed how the colonial state by introducing new land revenue system and private property in land initiated fundamental changes. Thereby the peasants not only lost their occupancy right but became more and more impoverished. On the other hand due to legal support of the colonial state the Zamindars became virtually the omnipotent authority. Chakraborty also discussed why agricultural productivity declined in Bengal by 20thcentury. Sahara Ahmed discussed about the forest policy of pre-colonial, colonial and the post 1947 government of India in the Northern parts of Bengal. She tried to show that the forest dwelling communities enjoyed an autonomy during Mughal rule which no longer remained under colonial rule. During colonial rule these communities lost their customary rights and also were dispossessed. The plight of these communities remained same after 1947. W. Jyotirmoy Singh portrayed the history of communication in the North

East India. He showed that in Pre-colonial period there existed trade routes in North East India through which people of the region traded with Myanmar, Bhutan etc. He argued that during colonial period due to expansion of tea plantation in Assam the state built up a railway network. The partition of 1947 heavily affected the communication in North-East and one solution to the problem is to open communication with Bangladesh.

The Second section is consisted of articles by Kaushik Chakraborty, Sujaya Sarkar and Ramendra Nath Bhowmick under the title *Ambiguity of culture*. Large number of printed books in English with a *genre* as well as historicism owing its origin in British cultural perception of India shaped by Shakespearean tradition were disseminated in India during colonial rule. This is the theme of Kaushik Chakraaborty's article. Penetration of printed English books and their subsequent adaptation in Bengali, Hindi, and Oriya etc. displaced the pre-colonial pre-print indigenous tradition of historicism, culture and learning. Instead, English civility and rationality paradigm became a yardstick and assumed a hegemonic form. Sujaya Sarkar has discussed about the ideals and activities of Moulana Abul Kalam Azad. Azad Gave utmost priority to secularism, cultural cosmopolitanism, spread of education, specially of women. Sarkar emphasized about the relevance of the ideals of Azad in the 21st century. Ramendra Nath Bhowmick discussed about the spread of school and college level education among the *adibasi* labourers working in the plantation industry of Duars region in North Bengal.

The third section is consisted of articles written by Sabyasachi Dasgupta, Manas Dutta and Pinaki Roy under the title *Army and the colonial legacy*. Sabyasachi Dasgupta discussed about the recruitment policy in the regiments of colonial army. Dasgupta argued that in comparison to

Bengal army, the Madras and the Bombay army applied a different policy of recruitment. Whereas Bengal army was composed of high caste Hindus, specially Brahmins, the Bombay and Madras army was composed of non-Hindus, muslims, low-castes along with high caste Hindus . What became visible in post-1947 India was a continuity and survival of the "ethos" of colonial period's army in the garb of valor. Manas Datta discussed about the role of lower caste Paraiyans in the Madras army during the colonial rule. Whereas in the early and middle part of the 19th century the Paraiyans were recruited in a large number in the Madras army but from late 19th century the "martial race" theory regarding the recruitment policy denied Paraiyans further service in the army. Pinaki Roy argued that the colonial state of India deployed a huge number of Indian soldiers both during the first and the Second World War in different battle fronts of Europe and other parts of the world. They were deployed on behalf of the British imperial power. Roy argued that while Indian soldiers contributed a lot, showed great courage but they were treated as inferior, subservient, kept ill equipped only because they belonged from a colonized background.

The final section is consisted of articles by Kartik Chandra Saha, Tapas K. Roychoudhury and Ashim Kumar Sarkar under the title of *Resistance*. Saha and Roychoudhury discussed about the earliest resistance against colonial rule, i.e., *Sannyasi-Fakir* uprising of Bengal (1760-1800). The article tried to portray the character of the uprising going beyond the conclusions reached by the existing scholarship. Saha and Roychoudhury raised question about the category of Sanyasi-Fakir, were they peasants or traders or resemble with something equivalent to class. The article highlighted about Zamindar, peasant and Sannyasi-Fakir mutual support in resisting the colonial state.

Notes :

1. Peter Robb, *Liberalism Modernity and the Nation*, New Delhi, 2007, p. 4.
2. Kaushik Chakraborty, *Decolonising the Revolt of 1857*, Kolkata, 2007, p-214.
3. James Mill , *History of British India*, Vol I & II, Ed. H. H. Wilson, London, 1858, p-113-117, Vol. I; p-114, p. 132, pp. 342-347, Vol II.
4. Ashim Kumar Sarkar and Kaushik Chakraborty (ed.), *New Aspects on Indian History*, Kolkata, 2011, p. VIII.
5. *Sambad Prabhakar*, 5 December 1840, BSP 168, Archive of CSSSC, Kolkata.
6. Bhudeb Mukhopadhyay, *Swapno addho Bharatbarsher Itihas*, Hugli, 1302 B.S., p. 57.
7. Kaushik Chakraborty, Decolonising the Revolt of 1857, *op.cit.*, pp. 223-224.
8. Ashim Kumar Sarkar and Kaushik Chakraborty (ed.), *op.cit.*, p. IX.
9. *Ibid.*
10. A. Raghuramaraju, *Modernity in India Social Theory*, New Delhi, 2011, p. 1.
11. *Ibid.*
12. *Ibid.*
13. Partha Chatterjee, 'Our modernity' in *Empire and Nation*, New Delhi, 2010, p. 146.
14. Dipesh Chakraborty, *Habitations of Modernity*, New Delhi, 2006, p. 39.
15. Partha Chatterjee, *op. cit.*, p. 152.

TRAJECTORIES OF DEVELOPMENT AGRICULTURE , ENVIRONMENT AND COMMUNICATION

Agrarian Relations and Agriculture in Bengal during British Era : A Critical Re-evaluations

Anjan Chakrabarti

Introduction :

The Bengal (undivided) was always identified as possessing the most fertile land and this was because it always used to be lying in the delta created by three great rivers the Ganges, the Brahmaputra and the Meghna. Adequate rainfall, surface water, under ground aquifers and warm year round temperature made Bengal an ideal region for the growth of agriculture. Early travelogues and various historical accounts reaffirmed this fact. Ibn Battuta, the Moroccan explorer while visiting Bengal in the middle 14thcentury asserted, 'this is a country of great extent, one in which rice is extremely abundant. Indeed I have seen no region of the earth where provisions are so plentiful'. [1] The French traveller Bernier who visited the Mughal Empire during the reign of fifth Mughal Emperor Shah Jahan being impressed, while viewing the uttered prosperity of Bengal : [2]

> 'Egypt has been represented in every age as the finest and most fruitful country in the world, and even our modern writers deny that there is any other land so peculiarly favoured by nature; but the knowledge I have acquired of *Bengale,* during two visits paid to that kingdom, inclines me to believe that the pre-

> eminence ascribed to Egypt is rather due to *Bengale*. The latter country produces rice in such abundance that it supplies not only the neighbouring but remote states....*Bengale* abounds likewise in sugar...the three or four sorts of vegetables which, together with rice and butter, form the chief food of the common people, are purchased for the merest trifle, and for a single *roupie* twenty or more good fowls may be bought. Geese and ducks are proportionately cheap. There are also goats and sheep in abundance....fish of every species, whether fresh or salt, is of same profusion. In a word, *Bengale* abounds with every necessary of life....There is in *Bengale* such a quantity of cotton and silks, that the kingdom may be called the common storehouse for those two kinds of merchandise, not of *Hindoustan*or the Empire of the *Great Mogol* only, but of the neighbouring kingdoms, and even of Euope'.

However, the moot point was the ownership of this fertile tract, and in the pre-British period the generally accepted view was that, 'the question whether the land was owned by the Raja, the *talukdar*, *zamindar*, the cultivator or the king was not a real question. Each had claim based upon custom or upon grants made by the king or a raja, or upon grants made by a *talukdar* or *zamindar*. But in addition to these claims each did or did not have the power to enforce his claims or to enlarge his claims. There was no appeal if a usurper had the power to usurp. The village did not hold its land common. Common were its officials and servants'.[3]

The European travellers who visited India during 16th and 17th centuries had the unanimity on the notion that 'king was the owner of the soil in India'. In the latter phase, East India Company and the British government had similar view and exerted the view that they had 'inherited universal right of ownership over the land from their predecessors'.[4]

On the question of agrarian relationand the condition of the peasantry share-cropping emerged as one of the important determinants. From the various accounts of sharecropping in India and Bengal, authors were unanimous that sharecropping was as old as the country itself. Old Buddhist texts and Hindu law books of the ancient period bore the testimony of sharecropping. In the 5th Century B.C. lands belonged to Buddhist-monastery were cultivated by the sharecroppers because Buddhist monks were not allowed to participate directly in cultivation. The produce was divided equally between monastery and sharecroppers. It was also mentioned in *sloka* 166, chapter IV of Manu that non-cultivating landlords leased out their land in lieu of share of produce. It was also mentioned in the 'Laws of Manu' that the share was one-sixth of the gross produce, and Manu noticed that the sixth might be raised to one fourth in time of war or other emergency.[5]

During the Mauryan period (from 321 B.C. to 297 B.C), cultivators were employed to cultivate the Crown or government farms on receipt of share of produce. If the cultivators provided seeds and bullocks they received half of the shares of the produce but if they offered only their physical labour then they used to receive one-fourth or one-fifth shares of the produce, [6] Kautīlya's *Arthaśāstra* also referred, about the prevalence of sharecropping in ancient India and he described the system as follows : Unframed lands could be ... assigned to those who had nothing to supply but their bodily labour. The seed would be deducted and the women folk of the assignee subject to work at milling the grain taken by the state as its portion. Apparently the representative of the state arranged even for tools and oxen in such cases. From Megasthenes's *Indica,* it is learnt that the Mauryan king Asoka reduced the state share of crops produced from one-sixth to one-eighth. In the later centuries, serfdom on lands was practiced during the rule of various Hindu chiefs and regional governors and serfs were liable to pay a portion of produce. In ancient Bengal, this kind of practice was visible in Vishnupur and Dinajpur.[7]

One of the major reasons of the British East India Company to pursue revenue collecting right in Bengal presidency was to establish a monopoly over the lucrative textile trade in Bengal and the highly productive fertile land of Bengal. The statement made by Lord Cornowallis, the then Governor-General of Bengal in 1790 corroborated this fact, he stated, 'we have, by a train of most fortunate events, obtained the dominion of one of the most fertile countries on the face of globe, with a population of mild and industrious inhabitants, perhaps equal to, if not exceeding in number, that of all other British possessions put together'.[8]

It requires a mention that East India Company got the Grant of *Diwani* (revenue collecting rights) of Bengal, Bihar and Orissa from the Mughal emperor Shah Alam II in 1765 and after Awadh was annexed in 1856, the present India, Pakistan and Bangladesh had become part of British territory.[9]

The importance of colonial period is paramount while dealing with agrarian economy of India, especially undivided Bengal. Questions those were raised and debated during colonial and post colonial period were many and few of those questions are listed below : How did the tenancy system evolved in Bengal under colonial rule? What were the forms of tenancy? What were the terms and conditions of tenancy contracts? How efficient the share tenancy was in Bengal during the colonial period? What was the impact of the then prevailing tenancy structure on the level of agricultural production and productivity? This paper would make an honest attempt to find plausible answers of these questions.

1. Evolution of Tenancy System in Bengal during Colonial Period

East India Company, at the initial phase worked as a trading concern by acknowledging the authority of the *Nawab* of Bengal and the Mughal emperor in Delhi. The Company inched to the pinnacle of political power by winning the famous battle of Plassey in 1757. After installing a series of client rulers, East India Company finally got the Grant of *Diwani* (revenue collecting rights) of Bengal, Bihar and Orissa from the Mughal emperor Shah Alam II in 1765 and their attempt to collect revenue brought considerable changes in agrarian relations in Bengal. Initially they tried to work with per-existing institutional forms. In pursuit to increase Company's de facto power, they took the initiatives to create a new administrative infrastructure. The primary aim of the Company was to increase the collection of land revenue and as land taxes were closely linked with land ownership, as a consequence, East India Company was prompted to bring in some structural changes in land revenue system, administrative as well as judicial system in Bengal. *Zamindars* collecting revenues for Nawabs of Bengal were identified as natural intermediaries by the Company for collecting revenue. *Zamidars* were a heterogeneous group. *Zamindars* having huge estates, military capacity and judicial and administrative responsibilities could be called 'feudatory chiefs' and others who were operating at smaller scale might be called 'landlord'. [10] If the *Zamindars* were placed at the top of agrarian hierarchy on the land collection of taxes might be rested in the hands of *jotedars* who enjoyed occupancy rights on their land at customary rents . [11] *Jotedars* were also powerful at the village level and used to have tenants at their disposal. [12] The Board of Directors of the Company first introduced a quinquennial settlement or Revenue Farm Contract on a 5-yearly basis in 1772 and then a decennial settlement in 1789. Revenue collection rights were given to highest bidder through auction for a period of 5 years. Economic theory suggests that revenue farming has two negative outcomes. First revenue farmers, in the short run shall try to extract as much surplus as possible during his lease. Second, revenue farmers get trapped into "Winner's Curse" of auction theory i.e. they overestimate the value of asset and shall often default. This proved to be true for the revenue farmers of Bengal.[13]

This system of 'time-honoured' practice of revenue farm contracts forced the revenue farmers to resort to rack-renting and forcible collection and as a result cultivating peasantry had to face unbearable burden. It was argued by the economic historians that bullion control measures adopted by the Government in England had put pressure on East India Company to increase its collection of land revenue from Bengal presidency consisting of Bengal, Bihar and Orissa. The company in India was bestowed with the responsibilities to finance the collection of locally manufactured products to Europe, and to meet the expanding military expenditure which occurred because of the continuous pursuits of the East India Company to annex and occupy more regions of this subcontinent which was known as the Investment Policy of the Company.

1.1 The Permanent Settlement and the Agrarian Bengal

These experiments however proved to be disastrous for the peasantry and Bengal had to face a famine in 1770 on the other. Rack-renting by the revenue collectors raised serious discontentment and unrest among the cultivators and the Sannyasi Revolt (1763-1800), Midnapore Revolt (1766-1783), Chakma Revolt (Chittagong, 1776-1787) bore the testimony for it. Therefore, the British Government reached to a conclusion that Bengal required a more durable settlement and in 1793 the Permanent Settlement was introduced in Bengal. Through Permanent Settlement, the British ruling power ascertained their supreme right over land. The *zamindars* or landlords were identified as agents of the State for collecting and depositing the annual revenue which was fixed by the State in perpetuity for each estate. In lieu, the *zamidars* were granted proprietary rights of the estates. They were also allowed to sell or buy estates and get unhindered rights to obtain rents from their tenants. Lord Cornwallis, on behalf of the British government had conceived the Permanent Settlement in anticipation that switching over to fixed revenue system would encourage the *zamindars,* which had the sole responsibility for rent collection, land improvements and managements, to produce more as it happened to English farmers in England.

Since the revenue payments were fixed, it would provide incentives to *zamindar* to maximise their income and in the process, accrued profit by the *zamindars* would be invested in agriculture. As a result, agricultural production and extension of cultivation would improve. Cornwallis asserted, 'Land property will acquire a value hitherto unknown in Hindustan and the large capital possessed by many of the natives in Calcutta which are now employed in usury or monopolising salt and other necessaries of life will be appropriated to the more useful purposes of purchasing and improving land', *zamindars* would turn into 'economical landlords and prudent trustees of the public interest'. [14] It would also be beneficial for the State because the settlement would guarantee a stable revenue and cost of collection of revenue would also be minimised (Cooper, 1988: 21). It was therefore become imperative to protect the property rights of the *zamindars* to have a steady flow of revenue. Therefore, a hierarchical system of judiciary was set up : Native Commissioners were given charge to hear suits for small sums at village level. *Diwani Adalat*, the civil court was set up at district level and at the apex of the system, *Sadar Diwani Adalat* was created. However, Indians were kept out of higher judiciary on the pretext that they are corrupt. It was further assumed by the colonial ruler that the coercion with the cultivators would be avoided since they were not directly collecting the revenue from the tillers of the land. Therefore, the peasant unrest would be minimised.

In Permanent Settlement it was conceived that a fixed rent would also be introduced for the cultivators but this had never become a reality. [15] Prior to the introduction of Permanent Settlement the categories of landlords existed in Bengal province were as follows : The Old Independent Chiefs like the Rajas of Coochbehar in parts of Assam, Tripura etc. and they maintained their ownership over their territories by paying tributes to the imperial power in Delhi. The old established landlords (bhuswamis) like Rajas of Rajshahi, Burdwan and Dinajpur and they were the de facto rulers of their estates and regions. Revenue collectors having hereditary rights collecting revenue appointed by Mughal rulers. Revenue farmers who were managing their own cultivation and collecting revenue simultaneously.

However, after the introduction of Permanent Settlement, various categories of landed gentry through a process of relegation and elevation were emerged as a single category called *zamindar*. All were granted perpetual, hereditary, proprietary rights in the soil, subject to the payment of permanently fixed revenue to the *ryots*. The revenue was fixed at ten-eleventh of the assets, i.e. annual gross rental and the remainder and in the process one-eleventh was left for the landlords or *zamindars*. A group of sturdy, rich and less enterprising peasant class with proprietary rights were emerged by replacing sizeable numbers of old *zamindar*s at the top who became the victim of Sun-set Law or Revenue Sale Law under Permanent Settlement and the ordinary *raiyat* at the bottom. An estimate observed that within two decades of the operation of the Permanent settlement about half of the *zamindari* lands of Bengal changed hands. [16] Old *zamindars* find it a difficult proposition of paying revenue on time because uncertainty in agriculture was quite high due to uncertainty of monsoon rain and hence uncertainty was creeping in paying rent on time. They defaulted and due to mounted pressure on the peasants they deserted in some regions. [17]

Prior to the introduction of Permanent Settlement the *zamindars* had enjoyed enormous administrative and judicial power but after the introduction of the Permanent Settlement, these traditional powers were seized. Thus *zamindars* by and large perceived the Permanent Settlement as bane. Gross indignation and insecurity crept in the minds of them. This made a heavy in the psyche of old *zamindars*. *Zamindars* never got attracted to invest their surplus for the improvements of their estates by investing in reclamation, irrigation, *taqavi* loans, storage, better live stock etc. because they considered that opportunity cost of investment in purchasing new estates under cultivation, money lending, purchasing government bonds or shares of Agency houses were much higher in comparison to improvements of agriculture which was fraught with uncertainty. For the *zamindars*, three preferred areas of investment were observed. First, money lending to tenants, merchants and shopkeepers was considered to bring highest return and it also allowed the landlords to exploit the tenants for their perceived indebtedness. Second, landlords preferred to invest in government bonds and shares of agency because these investments would assure a ready income and it would also ensure government patronage that enhance landlords' social prestige. Third, surplus capital was also invested in purchasing new estates because it would allow creating permanent tenure at fixed rent in lieu of handsome money or *selami*.[18]

Four types of land tenures were prevalent under British rule and classified as under:

i. permanently settled estates (revenue paying) which were created during the introduction of Permanent Settlement;
ii. Revenue-free estates created by the pre-British rulers and recognised and legitimised by the imperial rulers during the introduction of Permanent Settlement;
iii. Temporarily settled estates that were created by the state with private persons as proprietors and this category was created after the abolition of Permanent Settlement;
iv. Estates held directly by the Government (*Khas*) and these estates were created after the introduction of Permanent Settlement where in it was overtly announced that all lands belong to the state. The following table (1.1) provides an account of area under different categories of land tenures.[19]

Table-1.1

Types of Land Tenure and Area under different Tenurial System

Type	Number held (in thousand)	Area (in million acres)	Percentage of Total
Permanent Settled Estate (revenue paying)	94	37.37	80.7
Revenue-Free Estate	51	1.97	4.3
Temporary Settled Estates	4	3.34	7.2
Estates hold directly by Government (Khas)	4	1.65	3.6
Total	153	46.33	100.0

Source: *Government of Bengal (1945), Famine Inquiry Commission, Final Report, p.437*

For any future increase in the assets resulting from the extension of cultivation or other causes, zamindars were also assured to accrue additional benefits. Zamindars, unlike the expectation of the authors of permanent Settlement, failed to increase the output and the condition of the peasantry, rather to augment their income, they prefer transfer and divide their rights and as a result a chain of intermediaries or intermediate tenures, a large section of rent receivers were created between zamindars and cultivators. This process was termed as sub-infeudation. [20] A study of Bakarganj district mentioned that in early twentieth century there existed as many as twenty intermediaries between cultivators and proprietor and one hundred and sixty two terms were used to describe various forms of tenures and sub-tenure. As a result, the burden of rent for the actual cultivators increased manifold. As pointed out by Ratnalekha Ray 'a traditional mode of production controlled by a creditor class of village landlords persisted on the basis of a backward technology. Large-scale capitalist farming through increased capital inputs and improved technique, as envisaged by the farmers of the Permanent Settlement, did not develop in Bengal'. [21] Thus Permanent Settlement and its failure raised questions on three important unsettled issues: 'first, what level of coercion could the landlord use on tenants? Second, how much freedom did he have to raise rents? And finally under what conditions could he evict a tenant?'. [22]

1.2 The Post-Permanent Settlement Era and Bengal Tenancy Act

Corporal punishment and jailing of defaulting tenants was common in pre-colonial Bengal. [23] In the regulations of Permanent Settlement, the landlords were stripped off from this extra judicial power. Therefore, zamindars were constantly complaining of non realisation of rents from the tenants. In response, the East India company in their continuous pusiut of increase tax collection introduced the regulation VII of 1799 (known as Haptam) and Regulation V of 1812 (known as panjam) gave the landlords and intermediaries absolute and unrestricted authority to use all kinds of coercive forces against the cultivators (e.g. seizing the crops, personal property, arresting and harassing physically) for realising rent arrears. [24] Cultivators thus become subject to all kinds of oppression and forced eviction and land alienation became order of the day. A significant rise in the proportion of landless cultivators was observed in Bengal during this period. [25] Therefore, the pious intention of the colonial rulers of capitalistic development of agriculture, as it was the case for England, and minimisation of peasant exploitation and unrest, remained completely unfulfilled and unaddressed through Permanent Settlement. These facts were also accepted by the then Government of India and it was observed that the desired effectiveness of the Permanent Settlement was not 'supported by the experience of any civilised country' and under this settlement 'cultivator was rack-rented, impoverished and oppressed'. [26] According to Land Revenue Commission, [27] 'the Permanent Settlement imposed on the province an iron-framework which has had the impact of stifling the enterprise and initiative of all classes of people'.

The failure of permanent settlement put the government in serious financial strain because the administrative cost was rising every year while government revenue got stationary.

It was increasingly being felt by the colonial rulers due to fears of agrarian unrest and the failure of the Permanent Settlement to augment agricultural growth that some protection became imperative for the cultivators and power to use coercive forces by the zamindars and other intermediaries for realising rents should be contained and reduced. To accomplish these objectives, Rent Act of 1859 was passed by the Colonial government. This law restricted the landlords' power of raising rents arbitrarily and their power to evict the tenants at their will. The right to occupancy based on twelve years continuous possession of a plot of land was also ensured by the Rent Act of 1859. However, the protection against rent revision and forcible eviction was extended only for two types of raiyats, namely Paikashtraiyat and khud Kashtraiyat leaving numerous sharecroppers in the lurch. The act created conflict between occupancy tenants and zamindars. Brewing discontent among the peasant community culminated into an agitation and Pabna Riots forced the British Government to enact a broad-based tenancy legislation in 1885 which was known as Bengal Tenancy Act (B.T.A). The B.T.A limited rental increases to 12.5 per cent with no further increase allowed for 15 years. [28] The primary goal of this legislation was to plug-in the loopholes observed during Permanent Settlement and Rent Act of 1859.

It needs to be remembered that Rent Act of 1859 as well as Bengal Tenancy Act of 1885 were enacted after the Sepoy Mutiny of 1857. The political situation underwent a rapid change after mutiny. Dissatisfaction was widespread among various classes of society and people because of the adverse impact created by the British policies and more specifically land and revenue policies. The growing discontent was exploded in the form of Mutiny of 1857 that posed a serious threat to the newly established British rule in the country. [29] Further, in Joshi's word,' the Mutiny marked an end to the era of British change and innovation in the Indian institutional structure, specially the property structure'. [30]

B.T.A of 1885 defined the rights of the raiyats and the transfer of holdings also received legal protection. In addition, the earlier distinction between Paikashtraiyat and khud Kashtraiyat were also abolished and both became settled raiyats provided they held any land in a village for 12 years. Furthermore, this legislation added various categories of tenants to minimise the exploitation of tenants in the hand of landlords as well as various rent seeking intermediaries to check the increasing peasant unrests against the oppressive landlords and intermediaries, to arrest the problem of continuous sub-infudation in agrarian Bengal.

This Act also created special provision to ensure land rights for the tribal people residing in Bengal. It was mentioned in the section 89 of Bengal Tenancy Act that no ejectment of tribal people from their land would be allowed without an execution of decree from the civil court. This act also initiated a process of a record of rights and settlement of land revenue. The Bengal Tenancy Act not only put restriction on arbitrary increase of rents but also underlined the rate at which rent could be enhanced. It was mentioned in the act that for secured categories of raiyats, rents could not be enhanced more than 2 annas in the rupee (i.e. 12.5%). Bengal tenancy Act was further amended in 1928 and 1938. Some major amendments included: holdings of raiyats and under-raiyats were made freely transferable though pre-emption rights were granted to the landlords; occupancy rights were given to the under-raiyats for tenancy over twelve years and burden of rent would not exceed one-third of gross produce.[31] Though these legislations seemed to be pro-tenants, however, the plight of the sharecroppers got strangled in the legal ambiguity. It was stated in clause 3 (3) of Bengal Tenancy Act of 1885 that sharecroppers might or might not fall into the category 'tenant' and it was depending on whether the sharecropper 'hold the land and deliver a share of the produce on account of its use and occupation'. If the sharecroppers fulfilled this clause then only they would be recorded as raiyats or under-raiyats but this solely depended on how that clause would be interpreted.[32] It can be said that B.T.A had 'strengthened the position of substantial landholders in the village vis-à-vis the zamindari but still left the lowest tier of the agrarian hierarchy unprotected'.[33] Agriculture during colonial era remained solely dependent on tenant cultivation (Table-1.2) who continued to face different kind of oppression in the hand of landlords, intermediaries, money lenders. Inspite of the fact that share cropping was used as oppressive mechanism why did share-cropping more precisely the bargadari system rose in 19th century Bengal?

While replying to the questionnaire of Bengal Land Revenue Commission (1941), four reasons were cited by Radhakamal Mukherjee and they were as follows :[34]

"1. The middle-class in Bengal which do not resort to cultivation advance money to raiyats and purchase occupancy holdings resettling the raiyat on a produce rent for their convenience.

2. The increase is especially noticeable in such areas as Midnapore, Dacca, Faridpur and Tippera.

3. Some agricultural castes ape the manners of Bhadroloks and are refusing to drive the plough and have acquired the habit of employing adhiars as soon as they acquire more than 12 to 15 bighas (4 to 5 acres) of land. This is marked in West Bengal.

4. The increase of transfer of holdings to non-agriculturists, marked in the recent years of agricultural depression, has led to the rapid and persistent spread of produce-paying tenancies."

The Bengal Provincial Kisan Sabha supplemented some additional reasons in their memorandum to the Commission in support of the spread of sharecropping in Bengal. It was mentioned that"the landlord does this (encourages share-cropping) as it provides a very simple way of raising rents without recourse to courts." It was further added that " growth of a type of landlords- the money lender cum landlord...At the same time he is a petty trader in jute or wheat or rice, very often the village grain trader or he may be interested in transport...But he is not intending to cultivate the land himself ... Now to a landlord of this type the receipt of produce rents has every advantage; not only the highest rate well worth the extra trouble of assessing and collecting his half share, but and this is most important, this new type of landlord is already a trader in grain or jute. It will pay him, therefore, to take as rent the jute or grain grown by his share cropper rather than to take cash and then have to buy these raw materials for his business in the open market. In other words, he is rationing business vertically. Consequently we find amongst these new landlords it is the usual practice to have their land cultivated by bargadars."

Failures of various legislations introduced by the Colonial ruler failed to protect the economic and social plights of tenants in Bengal and as a result from early 19th Century share-cropper movements in varying nature were kept on occurring and got more and more intensified and widespread from the beginning of 20th century. If 1830s, the Faraizi (Muslim reformist) movements led by Titu Miyan marked the beginning of expression of economic and social tension in eastern Bengal, then tribal insurrection against economic oppression in Santhal Parganas was manifested in terms of protest which was commonly known as Santhal rebellion. Antagonism against planter's systems of indigo cultivation culminated in 'Blue Mutiny' and the movement continued from 1870 to 1880. In 1909, in a village of Jessore district Muslim sharecroppers refused to cultivate the land of Hindu-landlords on demand for two-third share of produce. With the rise in population pressure on land, absolute scarcity of fertile land, reclamation of forest and infertile land for cultivation, Share-cropper movement became more sporadic and widespread. The scenario did not change even after independence and partition of Bengal and Tebhaga movements of 1946-47 bore the testimony of it.

Table-1.2
Proportions of Agricultural Population: 'Landlord', 'Tenant', 'Labourer' (in percentage)]

Districts (East Bengal)	Landlords	Tenants	Labourers
Bakarganj	4.26	87.75	7.99
Bogra	1.99	89.24	8.77
Chittagong	9.55	72.53	17.92
Dacca	5.08	89.24	5.68
Faridpur	6.61	88.84	4.55
Khulna	5.73	85.68	8.59
Mymensingh	2.49	92.12	5.39
Noakhali	3.36	82.50	14.14
Rajsahi	4.60	84.18	11.22
Tippera	1.93	93.87	4.20
East Bengal	**4.52**	**86.75**	**8.73**
Districts (West & Central Bengal)			
Bankura	4.18	68.39	27.43
Birbhum	2.00	62.38	35.62
Burdwan	4.19	68.42	27.39
Hooghly	4.87	68.57	26.56
Howrah	8.89	61.06	30.05
Jessore	6.94	85.34	7.72
Malda	2.12	75.99	21.89
Midnapur	2.37	80.45	17.18
Murshidabad	4.15	68.99	26.86
Nadia	7.85	70.05	22.10
24 Parganas	4.39	76.19	19.42
West & Central Bengal	**4.72**	**71.44**	**23.84**

Source:Census of India 1921, Vol.5, Bengal, Pt 2 **[Quoted in Bose, Sugata (1986):31]**

2. Agriculture during British Rule: A Critical Review

It was also pointed out that average productivity declined after 1910-11 and even became lower than the Mughal period.[35] This bound to have a negative impact on agrarian community. It has adequately been explained that introduction of Permanent Settlement and subsequent changes in land settlements did not succeed to bring in the private investment for agricultural development and also failed to break the spiral of exploitation meted to the sharecroppers by the zamindars, jotedars, intermediaries of varied nature and money lenders. This adversely affected the agricultural production and productivity in Bengal.

What were the other possible causes of decline of agriculture in Bengal which once compared as best fertile strip on the Earth? A fleeting glance through the pages of various historical documents and studies allow us to identify various non-economic and economic factors responsible for the retardation of agriculture and worsening of the economic condition of the majority of the agrarian community in Bengal.

Accepting the fact that the whole of Bengal was an alluvial plane, the early 20th century Bengal could be divided in two parts- land of agricultural decline consisting of ecological zone of the old delta including West Bengal and Central Bengal and land of agricultural prosperity consisting the new delta comprising East Bengal and the Ganges-Brahmaputra Doab corresponding to North Bengal.[36]

2.1 The Land of Agricultural Prosperity Consisting of the New Delta comprising East Bengal and the Ganges-Brahmaputra Doab corresponding to North Bengal

Most of the districts in East Bengal were created by the active deltas of the Ganges and Brahmaputra. The area got flooded every year and as a consequence, the region received a fresh fertile layer of silt. The delta so formed by these two rivers could be divided in two sub-regions. Districts like Dacca, Mymensingh, Faridpur and Tippera were lying on the upper Ganges-Brahmaputra delta and the lower part of the delta was consisting of districts of Bakerganj, Noakhali, Khulna and Chittogong.

Reach alluvial tract of Dacca, Mymensingh, Faridpur and Tippera along with the Jamuna catchment area in Rangpur and Pabna and part of Rajsahi were suitable for multiple cropping. Aman rice and jute were two principal crops grown in this region. Abundant moisture and inundated plains of East Bengal helped to produce aman (one of the best quality grains among the traditional varieties of rice) in a large quantity. Two varieties of jute were generally found-capsularies which grew in low flooded land and oliotorious on higher land which produced a superior fibre but had to be manure in the absence of inundation silt. Other than aman and jute, sugarcane, pulses, oilseeds, wheat barley, oats and tobaccos were also grown in this region.[37]

2.2 The Land of Agricultural Decline Consisting of Ecological Zone of the Old Delta including West Bengal and Central Bengal

The West Bengal and the Central Bengal was the oldest agricultural settlement in India. Except the western part (Birbhum and Bankura) which contained laterite formation, rest of the West Bengal and Central Bengal was identified as 'semi-aquatic rice plain'.[38]

In fact, from the 17th century the rivers of the region started to change its course towards east and from 19th century the shift became faster. This caused the silting up of the rivers and simultaneous fall in sub-soil water level which made the West Bengal and Central Bengal a zone of agricultural decline.

Foothills of Himalayas as well as the districts of Malda, Dinajpur, part of Rajsahi and Bogra were also part of alluvial zone. The crops of old alluvial zone were aus rice, sugarcane, maize, joar, millets, pulses oilseeds, wheat, barley and jute in a small extent. It was also observed that the aus rice, whose yield level was relatively poor, grew on relatively high land and required less water than aman rice. As a consequence, we observed dominance of aus cultivation in alluvial zone of West Bengal where there was no fresh flow of water due to senility of the river system. The loss of silt-laden red water resulted in a switch to inferior crops and drastic fall in yields were observed.[39]

From the middle of 19th century, the natural drainage system was also severely affected with the construction of rail and road embankments. As a consequence, water got stagnated and became breeding ground of mosquitoes which resulted in series of malaria epidemics. People were dying regularly in the districts of Jessore, Hoogly, 24 Parganas, Nadia, Burdwan, Birbhum, and Howrah. This resulted in decline of rural population and serious decline in cultivated land (Table-1.3)was followed between 1891 and 1931.[40]

Table-1.3

Decline in cultivated land (in percentage) between 1891 and 1931in districts of West and Central Bengal

Districts	% decline
Burdwan	50
Hooghly	60
Jessore	30
Nadia	20
Murshidabad	20

Source: Mukherjee Radhakamal (1938), Changing face of Bengal: A Study of riverine economy: 275[41]

2.3 Declines in Agriculture in Bengal in 20th Century: Possible Causes and Consequences

However, 1920 and onwards, the growth and expansion of cultivation in East Bengal slowed down. Rapid population growth (Table: 1.4)was the main reason behind the agricultural slow down. Between 1920 and 1946, the 'Bengal agriculture had reached an equilibrium at a low levels of production shown by the near constancy of cultivated area and near zero-trend in yield per acre'. [42]

In 1919, the settlement officer of Mymensingh reported that cultivation had almost reached its full limits. In the Settlement Report of Dacca, in 1917, it was mentioned that the cultivated land as a percentage of cultivable land reached to 92 per cent.

Table-1.4
Density (Per Square mile) of Population in the districts of East Bengal in 193

Districts	Population Density
Dacca	1256
Tippera	1197
Noakhali	1124
Faridpur	1003
Bakarganj	834
Mymensingh	823
Pabna	795
Bogra	785
Rangpur	742

Source : Bose, Sugata (1986):47 [43]

Excessive population pressure in fertile land of East Bengal forced the people to move in search of new patches of land. Madhupur jungle located at the North-east of Dacca was reclaimed. Chunk of agrarian community started moving up towards the Jamuna and the Assam valley upto Gauhati and along the left bank of Brahmaputra as far as Tejpur.[44]

People from Bakerganj, Jessore, Khulna were started moving to Sundarban region in search of livelihoods.[45]

Between 1881-1901, land of less fertile barind area consisting of parts of districts of Malda, Bogra and Rajshahi was brought into cultivation and Santhals from Chota Nagpur area were engaged because of their experience of clearing jungle and forest land for cultivation .

Forest land of the western Duars (Jalpaiguri district) was also brought into cultivation. People migrated from other districts of North Bengal and Chota Nagpur and started working as coolies in the tea-gardens of duars or as share croppers.

Cultivation was even extended to less fertile laterite region of districts of Bankura and Birbhum and also in Tamluk and Contai sub-divisions of Midnapore. Therefore, with the intensification of population, cultivation was extended to lesser or least fertile zones. Holdings got further subdivided and became smaller in size. Therefore, slump in productivity and production was a natural logical outcome in Bengal.

It should further be noted that rise in population and adverse land-labour ratio accentuated the share-cropping on newly reclaimed land and competition for land 'forced cultivators to accept exploitative tenancy arrangement such as share contract'.[46]

Fluctuation of Price, especially for rice and jute had serious impact on Bengal's agriculture. We are well aware that during 19th Century, British mercantile pursuit encouraged them to produce cash crops like tea, cinchona, in the Himalayan region and its foot-hills, opium and indigo were also cultivated on khas lands of planters and cultivators were provided with advances which kept them permanently indebted. After 'indigo revolt' in 1860, its cultivation declined.[47]

Subsequently, by the end of 19th Century, jute was emerged as most important cash crop especially in East Bengal and rice and jute became driving force of Bengal's agrarian economy. Cultivations of rice did meet up not only consumption requirement but also supplemented the cash requirement through the selling of surplus production. Rising demand for rice both in domestic market and for export did contribute to the rise in production of rice between 19th and 20th Centuries.[48]

For jute cultivation, the required money for production was supplied by the traders in lieu jute cultivators hypothecated their crops to the traders. Hence, involvement in the world market also made the crops (especially jute and rice) susceptible to the fluctuations in the prices in the world market. Fall in price along with demand, as it reached to nadir during world wide depression of 1930s put the cultivators under serious financial stress and they became easy prey for the village scourges, the money lenders. As a matter of fact, between 1929 and 1933, prices of rice and jute were fallen by 68 per cent and 61 per cent respectively. Poorer families failed to grow enough rice to feed their family members and cultivated jute for sale. As the price of jute plummeted, poor cultivators were caught in debt trap and they became forced to sell their land and eventually either they became *bargadar* or else opted for bhag or adhi.[49] Since, cultivators of Bengal had little option of growing anything other than rice and jute which were equally remunerative like rice and jute, as a result they continued to languish between 'traditional form of cultivation and traditional form of exploitation'.[50] As Irfan Habib pointed out that jute was produced in Bengal during Mughal period but its manifold increase in cultivation took place during British period at the expense of rice and sugar cultivation. This attributed food shortage in Bengal during Colonial period.[51]

As per the report of the Bengal Banking Enquiry Committee 1929-30, average debt per family was from Rs 112 in 1928 to Rs. 187 in 1933 in spite of the fact that credit availability was very scanty.[52]

The Second World War of 1939 and subsequent rise in prices of rice, series of natural disaster between 1942 and 1943, lacklustre attitude of the British Government in intervening in market to control the price rise of foodgrains and hoarding of foodgrains by the unscrupulous traders resulted in famine in Bengal and three million people died in starvation. The worst example of man made famine was created in the history of human civilisation. The famine was the result not of shortage on the supply of food but lack of purchasing power in the hands of the poorer classes.[53]

Conclusion

From the foregoing analysis the assertion of Ibn Battuta about Bengal that he made on mid of 14th century that he had not seen a region on the earth where provisions are so plentiful sounds mythical and utopian. The British regime failed to bring any substantive changes in either in the methods of agricultural production or in the relations of production. Capitalist farming remained a distant reality. Neither efficiency nor the equity was ensured for the agrarian community in Bengal. Backwardness, poverty, increased social cleavage were some general outcome that colonial rule bestowed on Bengal. Overall agricultural condition even became worse in comparison to Mughal rule. Finally, with the independence of country and partition of Bengal crippled the agriculture and agrarian communities of both East Bengal and West Bengal.

The process of producing bulk of the jute in East Bengal and shipping it to Calcutta was permanently halted and the negative impact could be noticed even today. And the condition of Bengal's agriculture can be best explained in the words of Sir Francis Floud who headed Bengal Land revenue Commission of 1940 observed that instead of providing incentives for improved agricultural performance, the *zamindari* system was imposed and the proliferation of intermediate tenures 'has resulted in dissipating the responsibility for the best use of the land in the interest among a host of rent-receivers, all of whom have to be supported by the labour of cultivator, and none of whom have either the incentive or the power to exercise any control over the use of land. The Government has done far less to develop increased production from the land than the Governments of other provinces'. [54] British explanation on rural poverty and backwardness in India had bypassed the agrarian structure created by them, as Myrdal did in his exposition of 'colonial theory' of poverty and economic backwardness. Colonial theory was aptly summarised by Nehru and he said 'If India is poor, that is the fault of her social customs, her *banias*, and money-lenders and above all her enormous population.' [55]

British-led colonialism had accentuated the de-industrialisation process in Bengal, as a result, population pressure kept on increasing on land. Landlords, money lenders and traders appropriated surplus from agriculture but accumulated capital in terms of surplus was not invested either on agriculture or in industry. Colonial rulers went for commercialisation of agriculture without bringing any concomitant technological changes in agriculture; this added further uncertainty to the peasantry having small land holdings and agricultural backwardness continued.

References :

B.H. Baden-Powell, (1876), *The Indian Village Community: examine with reference to the physical, ethnographical and historical conditions of the provinces; chiefly on the basis of the revenue settlement records and district manuals*, Longmans, Green & Company, New York.

Baden-Powell, B.H. (1892), *Land Systems of British India*, 2 vols, Clarendon Press, Oxford.

Bandopadhyay, N. and N. Krishnaji. (1983), *Evaluation of Operation Barga and the Impact of Land Ceiling Legislation undertaken in West Bengal*, Centre for Studies in Social sciences, Calcutta.

Bernier, F. (1914), *Travels in the Mogul Empire, AD 1656-1668* (2nd edn), revised by V. Smith, London: Humphrey Milford and Oxford University Press.

Bhatia, B.M. (1967), *Famine in India 1860-1965*, Asia Publishing House, Bombay.

Blyn, George. (1966), *Agricultural Trends in India, 1891-1947: Output, Availability, and Productivity*, University of Pennsylvania Press, Philadelphia.

Bose, Sugata. (1987), *Agrarian Bengal Economy, Social Structure and Politics, 1919-1947*, Cambridge University Press: Cambridge.

Bose, Sugata. (1993), *Peasant Labour and Colonial Capital: Rural Bengal since 1770*, Cambridge University Press: Cambridge.

Byres, T.J. (1983), 'Historical Perspective on Sharecropping', *Journal of Peasant Studies*, Vol.10:2.

Chaudhuri, B.B. (1970), 'Growth of Commercial Agriculture in Bengal 1859-1885', *Indian Economy and Social History Review*, Vol.7: 25-60.

Chaudhuri, B.B. (1970), 'Growth of Commercial Agriculture and its Impact on Peasant Economy', *Indian Economy and Social History Review*, Vol.7: 211-252.

Chowdhury-Zilly, A.N. (1982), *The Vagrant Peasant: Agrarian Distress and Desertion in Bengal 1770 to1830*, Franz Steiner Verlag, Wiesbaden.

Cooper, Adrienne. (1988), *Sharecropping and sharecroppers' Struggles in Bengal 1930-1950*, K.P. Bagchi& Company, Calcutta.

Firminger, W.K. (ed.) (1917), *The Fifth Report of the Select Committee of the House of Commons on the Affairs of the East India Company, Dated 28 July 1812*, Cambray& Company, Calcutta.

Government of Bengal (1940), *Bengal Land Revenue Enquiry Commission Report*, Vol.I, Calcutta.

Government of East Bengal and Assam (1909), Dept. of Revenue, Proceedings of the Conference of Commissioners, Shillong.

Guha, Ranajit (1963), *A Rule of Property for Bengal: An Essay on the Idea of the Permanent Settlement*, Mouton & Company, Paris.

Habib, Irfan. (1999), *Agrarian System in Mughal India 1556-1707*, Second Revised Edition, Oxford University Press, New Delhi.

Islam, M.M. (1978), *Bengal Agriculture 1920-1946 A Quantitative Study*, Cambridge University Press, Cambridge.

Islam, Sirajul. (1979), *The Permanent Settlement in Bengal: A Study of its Operation 1790-1819*, Bangla Academy: Dacca.

Islam, Sirajul. (2009), 'Bengal Agrarian Society Continuity and Change under the Colonial Rule' in ChittabrataPalit (ed.) *Agrarian Bengal Under the Raj*, SetuPrakashani, Kolkata: 37-60.

Joshi, P.C. (1957), *Rebellion 1857; A Symposium*, People's Publishing House, New Delhi.

Joshi, P.C. (1975), *A Survey of Research in Economics*, Volume Four, Agriculture-Part-II, Allied Publisher, Bombay.

Kling B.B. (1966), *The Blue Mutiny, Indigo Disturbances in Bengal, 1859-1862*, University of Philadelphia, University of Pennsylvania Press.

Klein Ira. (1972), 'Malaria and Mortality in Bengal 1840-1921', *Indian Economy and Social History Review*, Vol.9 (2): 132-60.

Kosambi, D.D. (1970), *The Culture and Civilisation in India in Historical Outline*, Vikas Publications, New Delhi.

Mclane, J.R. (1977), 'Revenue Farming and the *Zamindari* system in Eighteenth-Century Bengal' in R.E. Frykenberg (ed.) *Land Tenure and Peasant in South Asia*, Orient Longman, New Delhi.

Mukherjee, Radhakamal. (1938), *Changing face of Bengal: A Study of Riverine Economy*, University of Calcutta, Calcutta.

Mukherjee, Radhakamal. (1940), 'Indian Land System Ancient, Medieval and Modern, with Special Reference to Bengal' in Government of Bengal, Bengal Land Revenue and Enquiry Commission Report, Vol. 11, Bengal Government Press, Calcutta.

Myrdal, Gunnar. (1970), *The Challenge of World Poverty*, Penguin, London.

Norman Dorothy (1965), *Nehru: The First Sixty Years*, Vol.II, Asia Publishing House, Bombay.

Neale, W. (1969), 'Land is to Rule', in R.E. Frykenberg (ed.), *Land Control and Social Structure in Indian History*, Reprint (1979), Manohar, Delhi.

O'Malley, L.S.S. (1917), *Bengal,Bihar, Orissa and Sikkim*, Cambridge University Press, Cambridge, U.K.

Ray, Ratnalekha (1979), *Change in Bengal Agrarian Society, c1760-1850*, Manohar, New Delhi: 285-286.

Rothermund, D. (1978), *Government, Landlord and Peasant in India*, Franz Steiner Verlag, Wiesbaden.

Roy, Tirthankar. (2000), *The Economic History of India 1857-1947*, Oxford University Press, New Delhi.

Sen A.K. (1981), *Poverty and Famine: An Essay in Exchange Entitlements and Deprivation*, Clarendon Press, Oxford.

Sinha, N.K. (1962), *The Economic History of Bengal: From Plassey to the Permanent Settlement*, Vol. II, Firma KLM, Calcutta.

Swamy, Anand V. (2011), 'Land and Law in Colonial India' in Debin Ma and Jan Luiten Van Zanden (eds.) *Law and Long-Term Economic Change A Eurasian Perspective*, Standford University Press, Standford, California: 138-157.

Yule, Col. H. (1866), *Cathay and the Way Thither*, London: Hakluyt Society.

Notes :

1. Yule, Col. H. (1866), *Cathay and the Way Thither*, London: Hakluyt Society, p-257.
2. Bernier, F. (1914), *Travels in the Mogul Empire, AD 1656-1668* (2[nd]edn), revised by V. Smith, London: Humphrey Milford and Oxford University Press, pp. 437-39.
3. Neale, W. (1969), 'Land is to Rule', in R.E. Frykenberg (ed.), *Land Control and Social Structure in Indian History*, Reprint (1979), Delhi, pp. 3-15.
4. Habib, Irfan. (1999), *Agrarian System of Mughal India 1556-1707*, Second Revised Edition, Oxford University Press, New Delhi, pp. 29, 35.
5. B.H. Baden-Powell, (1876), *The Indian Village Community: examine with reference to the physical, ethnographical and historical conditions of the provinces; chiefly on the basis of the revenue settlement records and district manuals*, Longmans, Green & Company, New York.
6. Kosambi, D.D. (1970), *The Culture and Civilisation in India in Historical Outline*, Vikas Publications, New Delhi, pp. 150, 152.
7. Mukherjee, Radhakamal. (1940), 'Indian Land System Ancient, Medieval and Modern, with Special Reference to Bengal' in Government of Bengal, Bengal Land Revenue and Enquiry Commission Report, Vol. 11, Bengal Government Press, Calcutta, p-136; Byres, T.J. (1983), 'Historical Perspective on Sharecropping', *Journal of Peasant Studies*, Vol. 10(2), p. 7.

8. Firminger, W.K. (ed.) (1917), *The Fifth Report of the Select Committee of the House of Commons on the Affairs of the East India Company, Dated 28 July 1812*, Cambray& Company, Calcutta, p. 542.
9. Roy, Tirthankar. (2000), *The Economic History of India 1857-1947*, Oxford University Press, New Delhi, p. 1.
10. Bose, Sugata. (1993), *Peasant Labour and Colonial Capital: Rural Bengal since 1770*, Cambridge University Press: Cambridge, p. 70.
11. Sinha, N.K. (1962), *The Economic History of Bengal: From Plassey to the Permanent Settlement*, Vol. II, Firma KLM, Calcutta, pp. 18, 27.
12. Ray, Ratnalekha (1979), *Change in Bengal Agrarian Society, c1760-1850*, Manohar, New Delhi, pp. 6-7.
13. Swamy, Anand V. (2011), 'Land and Law in Colonial India' in Debin Ma and Jan Luiten Van Zanden (eds.) *Law and Long-Term Economic Change A Eurasian Perspective*, Standford University Press, Standford, California, pp. 141, 142.
14. Guha, Ranajit (1963), *A Rule of Property for Bengal: An Essay on the Idea of the Permanent Settlement*, Mouton & Company, Paris, pp.172, 173.
15. Bandopadhyay, N. and N. Krishnaji. (1983),*Evaluation of Operation Barga and the Impact of Land Ceiling Legislation undertaken in West Bengal*, Centre for Studies in Social sciences, Calcutta, p. 4.
16. Islam, Sirajul (1979), *The Permanent Settlement in Bengal: A Study of its Operation 1790-1819*, Bangla Academy, Dacca.
17. Chowdhury-Zilly, A.N. (1982), *The Vagrant Peasant: Agrarian Distress and Desertion in Bengal 1770 to1830*, Franz Steiner Verlag, Wiesbaden, chapter-V.
18. Islam, Sirajul. (2009), 'Bengal Agrarian Society Continuity and Change under the Colonial Rule' in ChittabrataPalit (ed.) *Agrarian Bengal Under the Raj*, Setu Prakashani, Kolkata: 37-60, p. 47.
19. Government of Bengal (1945), Famine Inquiry Commission, Final Report, p. 437.
20. Raychaudhury, Tapan (1969), 'Permanent Settlement in Operation: Bakarganj District, East Bengal' in R.E. Frykenburg (ed.) Land Control and Social Structure in Indian History, Madison, Wisconsin.p.167 ;Guha, Ranajit (1963), *A Rule of Property for Bengal: An Essay on the Idea of the Permanent Settlement*, Mouton & Company, Paris, pp. 172-173.
21. Ray, Ratnalekha, *op.cit*, pp. 285-286.
22. Swamy, Anand V., *op.cit.*, p. 146.
23. Mclane, J.R. (1977), 'Revenue Farming and the *Zamindari* system in Eighteenth-Century Bengal' in R.E. Frykenberg (ed.) *Land Tenure and Peasant in South Asia*, Orient Longman, New Delhi, p. 24.

24. Bandopadhyay and Krishnaji, *op.cit.*, p. 5.
25. Bandopadhyay and Krishnaji, *ibid*, p. 5.
26. Government of India, (1902), 'Land Revenue Policy of India', Department of Revenue and Agriculture, India, p.8.
27. Government of Bengal (1940), *Bengal Land Revenue Enquiry Commission Report*, Vol. I, Calcutta, pp. 35-36.
28. Rothermund, D. (1978), *Government, Landlord and Peasant in India*, Franz Steiner Verlag, Wiesbaden, p. 104.
29. Joshi, P.C. (1957), *Rebellion 1857; A Symposium*, People's Publishing House, New Delhi.
30. Joshi, P.C. (1975), *A Survey of Research in Economics*, Volume Four, Agriculture-Part-II, Allied Publisher, Bombay, p. 6.
31. Government of Bengal, *op.cit.*, pp. 25-30.
32. Government of East Bengal and Assam, 1909 (Report).
33. Swamy, Anand V., *op.cit.*, p. 148.
34. Mukherjee, Radhakamal. (1940), 'Indian Land System Ancient, Medieval and Modern, with Special Reference to Bengal' in *Government of Bengal, Bengal Land Revenue and Enquiry Commission Report*, Vol. 11, Bengal Government Press, Calcutta.
35. Habib, op.cit., pp. 21-23.
36. O'Malley, L.S.S. (1917), *Bengal,Bihar, Orissa and Sikkim*, Cambridge University Press, Cambridge, pp. 3-16 ; Mukherjee, Radhakamal. (1938), *Changing face of Bengal: A Study of Riverine Economy*, University of Calcutta, Calcutta , pp. 110-113.
37. Mukherjee, *ibid.*, p. 43.
38. Bose, Sugata. (1987), *Agrarian Bengal Economy, Social Structure and Politics, 1919-1947*, Cambridge University Press: Cambridge, pp. 37-38.
39. Mukherjee, Radhakamal. (1940) , *op.cit.*, pp. 83-84.
40. Chaudhuri, B.B. (1970), 'Growth of Commercial Agriculture in Bengal 1859-1885', *Indian Economic and Social History Review*, Vol. 7, pp. 25-60; Klein Ira. (1972), 'Malaria and Mortality in Bengal 1840-1921', *Indian Economic and Social History Review*, Vol.9 (2): 132-60, pp. 132-60.
41. Mukherjee Radhakamal, Changing Face of Bengal, *op.cit.*, p. 275.
42. Islam, M.M. (1978), *Bengal Agriculture 1920-1946 A Quantitative Study*, Cambridge University Press, Cambridge, p. 203; Blyn, George. (1966), *Agricultural Trends in India, 1891-1947: Output, Availability, and Productivity*, University of Pennsylvania Press, Philadelphia, pp. 20-27.
43. Bose, *op.cit.*, p.47.

44. Mukherjee, Radhakamal. (1938), *Changing face of Bengal: A Study of Riverine Economy*, University of Calcutta, Calcutta , p. 275 .
45. Cooper Adrienne, *Sharecropping and Sharecroppers' Struggles in Bengal*, Kolkata, 1988, p. 27.
46. *ibid.*, p. 28.
47. Kling B.B. (1966), *The Blue Mutiny, Indigo Disturbances in Bengal, 1859-1862*, University of Philadelphia, University of Pennsylvania Press, pp. 200-213.
48. Bhatia, B.M. (1967), *Famine in India 1860-1965*, Asia Publishing House, Bombay, p. 8.
49. Cooper, Adrienne, *op. cit.*, pp. 37, 42.
50. *ibid.*, p. 29.
51. Habib, *op.cit.*, pp. 21-35.
52. *ibid.*, p. 40.
53. Sen A.K. (1981), *Poverty and Famine: An Essay in Exchange Entitlements and Deprivation*, Clarendon Press, Oxford, Chapter 6.
54. Government of Bengal (1940), *Bengal Land Revenue Enquiry Commission Report*, Vol. I, Calcutta, pp. 25-37.
55. Norman Dorothy (1965), *Nehru: The First Sixty Years*, Vol.II, Asia Publishing House, Bombay, p. 556.

Forest Policies and Historical Legacies : Participatory Forest Management and Development in Jalpaiguri, c. 1720s-1980s

Sahara Ahmed

The term 'Participatory Forest Management' has gained primacy in the recent past and has garnered support from policy makers, development practitioners and numerous grass root organizations. They propagate that to play its historical role the agenda of "development" should be based on participation. As a corollary, participation is advocated as the only way to save development from degenerating into a bureaucratized lopsided, dependency creating institution. Hence, participation is perceived as a means to empower the browbeaten section of the society, to augment their control over the resources and to engage them in/under their own care. The term 'participation' is evocative of commonness and group activities or gatherings to fulfill a common purpose. The dictionary (Oxford English) meaning of participation is "the action or fact of partaking, having or forming a part of it, joining in, engaging in, contributing to, to be associated with, to be involved in, to be sharing in or having something to do with". Accordingly it could have myriad connotations. It could either be transitive or intransitive; either mandatory or voluntary; either manipulative or spontaneous; either permanent or ephemeral. Participatory Forest Management has generated considerable interest and has been bandied amongst various sections of the society in an endeavour to settle both the stakes of the state and the people in the forests amicably.

Participatory Forest Management or Joint Forest Management (JFM), as it is popularly known in India, has been in vogue since the 1970s. However, if the various connotations of the term 'participation' are taken cognizance of then one could raise numerous queries with regard to the prevalence of participation in varied forms in different historical periods. This paper will be an endeavour to trace the evolution of participatory forest management as a historical process from the Mughal period to the post-colonial period culminating in the genre as of a development discourse in the context of the Jalpaiguri district in North Bengal. In the process it shall make an attempt to comprehend nuances of participation in forest management in terms of modern day implications.

Revenue interests had gained primacy since pre-colonial times and often impinged upon the forest tracts. However, the degree of forest depletion and subsequent agricultural expansion varied from place to place and they did differ significantly from pre-colonial to colonial regimes. If the Mughal 'system' in its process of expansion gradually created the social and economic context within which it functioned, it was in turn fundamentally affected by the 'others' whom it sought at least to dominate, if not to integrate. [1] These were the tribal populace who were unlikely to have become completely irrelevant with the crystallization of the medieval village community and nor were their economic activities absolutely unrelated to the interests of the Mughal 'agrarian system'. [2] On the contrary, the village communities of Mughal India probably lived in reasonable proximity to such societies. It is, in fact, quite likely that the Mughal 'agrarian system' actually required them in order to replenish its ever-growing agricultural labour force.[3]

Large portions of the empire, though uncultivated were, however, unlikely to have been totally devoid of human population.[4] Many such areas were in all probability, inhabited by societies for whom settled agriculture had not yet become a primary means of livelihood. Moreover, as one scholar opined that even within the cultivated area there existed not only variations of the 'agrarian system' but also socio-economic systems that were remarkably different from it.[5] Such seems to have been the case with the forest dwellers of North Bengal. In fact, the Mughal control over their territory was probably of a nominal nature, and as Moosvi has argued that the arazi figures provided for it in the Ain-i-Akbari were not calculated on the basis of any survey.[6] Chetan Singh has opined that, the need to provide these figures, in the first place, possibly arose from Abul Fazl's attempt to make social diversities appear in conformity with the system that had evolved, particularly in areas that were largely under the close administrative control of the Mughals.[7] Even tribal territories, which may have been only marginally agricultural were, therefore, represented as areas which were properly measured, regularly assessed and in no way different from those regions of the empire that were indeed agriculturally developed.[8]

Establishment of Mughal authority over North Bengal proved to be an arduous task. It was accompanied by several major peasant uprisings and was supported by the Cooch Behar nobility.[9] Bengal had always displayed its recalcitrance against the establishment of Mughal suzerainty taking advantage of its distance from Delhi. The Western Dooars or the Bhutanese Dooars, was the jungle tract in the foothills of the Himalayas. Prior to British annexation, the Dooars was under Bhutanese subjugation and constituted part of the Cooch Behar kingdom. This region was conquered by the Mughals in the early seventeenth century, but apparently never fully integrated into the state structure of the empire. [10] In case of the Dooars, the forest itself may have been a natural or ecological border for the expansion of the agriculturally based Mughals. [11] It was also probable that the sparsely populated jungle tract created a level of disinterest in them. A younger branch of the Cooch Behar Raj established the Baikunthpur Raj in Baikunthpur to the north-west of the present town of Jalpaiguri and earned the epithet of Raikat of Baikunthapur. After Mughal emperor Aurangzeb's accession to the throne the Subahdar of Subah Bangla, Mirjumla Khan was sent to subdue the king of Kamtapur-Kamarupa. In the final truce of 1711, the Mughals restored Cooch Behar as an independent Kingdom. It was then only a shadow of its former glory under the great Koch Kingdom which during its heyday in the sixteenth century had controlled Northern Bengal and large parts of Assam.

According to the terms of the treaty the chaklas of Karjinhat, Kakina and Fatehpur became a part of the Mughal Subah of Bengal, under the faujdar of Rangpur. [12] The faujdar gave the chaklas of Boda, Patgram and Purbabhag in ijara to Chatranazir Santanarayan to be enjoyed by him and his descendants. [13] Baikunthapur's fate remained enigmatic. According to a source it was ceded to the Mughals while others maintained that it remained an independent territory. [14] With the dwindling prospects of the Cooch Behar kingdom the Raikat-raj of Baikunthapur started vying for superiority in the region.

Scattered references indicated that societies which were predominantly non-agricultural in nature were viewed as primitive and as a threat to the settled agrarian areas where Mughal land revenue were methodically applied. [15] Quite clearly, therefore, these peripheral societies were considered by the Mughals as external to their own society. [16] Yet, because they were seen as an occasional hindrance to Mughal hegemony, the imperial authorities intermittently endeavoured to establish military dominance over these people as a prelude to more fundamental economic inroads into their territories. [17] In fact, what seems to have insulated them from Mughal authority, irrespective of distance, were the densely forested areas which usually surrounded, or were interspersed with, their territories. [18] However, in the context of the Jalpaiguri district, one has to be cautious about overemphasizing the autonomy of the pastoral tribes. They were not totally isolated socio-economic entities, and their distinctiveness from agrarian society did not prevent them from interacting with it. Two distinct patterns of resource utilization could be discerned in the Dooars. In the northernmost part of the region, inhabited mainly by the indigenous Meches and Garos who practiced shifting cultivation. Along a narrow strip of land in the southern parts, which bordered on the northern fringes of Rangpur, the Rajbanshis had advanced quite far in the direction of settled cultivation. In the intermediate region there was a mixture of both systems. Apart from these the Dooars was also used by migratory cattle-breeders of the adjoining districts. [19]

The indigenous population of the Dooars were not confined to a life of splendid isolation in their pristine ecological zones in the pre-colonial period. On the contrary the fact that they had their own special linkages with their neighbours was clear from the nature of trade and exchange in the region. The southern parts of the Dooars had long-distance trade contacts with other parts of Bengal. [20] After the harvest was reaped the plains people assembled with their bullocks under their leader called the dafadar, and travelled to the northern parts with rice and other allied products and bartered them for cotton, oranges and venison brought in by the Meches. [21]

Political developments induced a shift in the management of forests, which invariably affected the history of the forest tract of the present Jalpaiguri district. The forests comprised two compact blocks, viz., the Garodhat and Patlakhawa reserves, elsewhere it existed in patches. Annual firing of undergrowth (weeds, tallgrass) was induced in the two reserves to initiate the growth of grasses with the idea of creating shooting reserves for the Maharaja of Cooch Behar. Besides, the Maharaja of Cooch Behar who was by far the wealthiest land owner, the Raikat of Baikunthapur (descendant of the younger branch of the Cooch Behar family) owned about fifty to sixty miles of dense forests (mainly sal and sissu) known as the Baikunthapur Jungle Mahal estate. [22] Sal and Sissu were felled for exportation and some of the autochthons (Meches, Rabhas, Totos and others) were employed in these ventures. No timber was cut except when commissioned and value was always paid in advance. [23]

Bankatai (the clearing of forests) and subsequent settling of cultivators were adequate reasons for staking a claim to the zamindari of an area, a claim which imperial authorities readily recognized. [24] Given the incentives that were provided by the Mughals, it was very likely that many zamindaris emerged at the frontiers of cultivation as a result of the clearing of forests organized by entrepreneurs of this kind. [25] The actual work of felling was done by groups of professional woodcutters and there probably existed in some regions, entire castes which specialized in this kind of work.[26] Hence, adjacent to the core agricultural areas of the Mughal empire was a territory where a perpetual struggle went on between cultivators and forests.[27] These were also areas where different socio-economic systems interacted with each other.[28] It is probable that they were driven to this as much by the logic of an expanding agrarian economy as by the pastoralists' or forest-dwellers' need for essential commodities from settled agriculturists.

It can be inferred that many peripheral societies which had earlier been only marginally affected by the Mughal state were gradually drawn into playing an increasingly important role in its economy as well as its political fortunes.[29] These societies as Chetan Singh argues, were neither numerically nor militarily insignificant, and their active participation, either as fuller constituents of the empire or as its opponents, could hardly have been without consequence.[30]

One could elaborate that these forest communities participated in forest clearance operations and induced the growth of grass by preparing the ground for the creation of hunting reserves. They were also recruited for timber operations frequently.[31] As the Conservator of Forests pointed out that the forests though natural were not "virgin" forests indicated they had previously been worked.[32] Francis Buchanan-Hamilton's report confirms that the jungles of Duars were not virgin forest at the time of British annexation. His report purveys well organized timber operations in Northern Bengal. He corroborated that the assimilation of labour gangs (indigenous population) in organized timber operations was observed, trees were cut and floated on the Teesta River and later sold to timber merchants.[33] These forest communities were also engaged in cultivating wastelands. Hence, participation of forest communities in forestry operations existed in the pre-colonial period though the form might have differed from its colonial or post-colonial successor.

The establishment of the Indian Forest Service, as per the Government Forest Act of 1865, was the harbinger of systematic and efficient planning in forest management. It is a conjecture that the phenomenal timber demand of the railway network had triggered the need for conservation. In fact, the state's forest policy evolved over time with its changing needs. The appointment of the first "Conservator of Forest" heralded a new regime in the forests of Bengal. The focus was on extraction and plantation of particular "valuable trees". In Dooars, this was above all the sal tree and later, through the plantations, both the sal and teak tree. Colonial documents are replete with references to a sort of underlying assumption that absolute state control was motivated by the need for protecting the forest against unwarranted destruction by local people besides private, commercial interests. Despite such concerns, the true outcome of British penetration of the forest tract was an age of unparalleled destruction.

In the colonial perspective, swidden or shifting cultivation was fraught with devastating consequences for the forests and arguments were raised about its negative climatic impact, causing distruption of rainfall and groundwater levels. It was reduced to "a social evil, the eradication of which would benefit the shifting cultivators themselves". [34] The demarcation of all the reserves in the Jalpaiguri district was completed in 1876-77. Large tracts of forest declared as 'wastelands' were opened up for tea cultivation and leased out to European planters at very nominal rents. The sal bearing forest tracts were demarcated and all jhum or temporary cultivation was prohibited there. As the government spread its administrative tentacles a systematic erosion of customary rights was witnessed with regard to collection of usufruct and the marginalization of the forest dwellers followed in terms of eviction.

In the meanwhile, Herculean efforts were made to raise plantations of certain specific species. Baffled by the problems of natural regeneration, plantations were raised in a planned manner. In the Terai and the Duars, the earliest plantation (1867) was aimed at introducing teak and mahogany near Sukna. This, however, ended in a faux pas on account of the poor quality of seeds and also due to poor drainage inherent in the site chosen for the trial. Denudation of forest resources and rearing of plantations exclusively for commercial exploitation had brought home to the colonial authority at least one lesson, that plantations cannot survive in the absence of undergrowth and other miscellaneous species. Besides, it was also clear that the clearance of undergrowth to prevent fires had rendered the young sal crop susceptible to crop disease Loranthus. [35] Despite the forest department's ability to control fires a new problem emerged : the sal crop stopped regenerating itself. The reduction of forest fires caused a tremendous spread of creepers, and even worse, the savannah grass cover favourable to sal seedlings was replaced by a thick layer of undergrowth, in which the young sal plants did not survive.

The dispossessed forest-dwellers expressed their dissatisfaction through sporadic revolts, sabotage in the form of forest burning. However, by 1900, the colonial government realized the significance of Taungya cultivation (i.e., cultivation of agricultural crop in conjunction with forest tree species) and the induction of village labour in fire protection and felling operations inaddition to other minor works. A constant supply of labour in fire protection and felling operations in addition to other minor works. A constant supply of labour for forest operations was deemed essential by the foresters hence forest villages were created. A half-hearted compromise was noticed in implementing taungya cultivation. Despite this, the British Government failed to implement a policy which would induce full co-operation from the villagers due to the government's notions of absolute property which often proved oppressive. Excessive control exercised by the British local potentates often forced the forest villagers to flee. The crops obtained from the taungya plantations proved insufficient to maintain a household throughout the year. It was also a risky type of agriculture, both the villagers' own crops and saplings often perished due to the depredations of wild pigs, elephants or other wild animals. The crops obtained from the taungya plantations proved insufficient to maintain a household throughout the year. It was also a risky type of agriculture, both the villagers' own crops and the saplings often perished due to the depredations of wild pigs, elephants or other wild animals. The colonial Forest Department, therefore, had to provide for the forest villagers with additional land for permanent cultivation instead of shifting cultivation. To minimize the loss of forest area, the Forest Department tried to select low-lying land, which is considered less suitable for tree plantations as the villagers' permanent plots. Each family received a document stating which plots they were entitled to cultivate. M.K. Raha pointed out that, "the land, the house and other constructions are not their own, but actually belong to the Forest Department of the State Government. The forest authority only permits them to enjoy these temporarily in exchange of their labour rendered to this department on year to year to contract." [36]

In 1946 after his retirement, Shebbeare made a final visit to the Dooars and pointed out that the taungya system that had been made possible, owed much to the Rabhas. He further stated that among the Rabhas one could find more skilled foresters than among officials like himself, but even so the Rabhas had received little in return. It was, he said, "about time that we did something to improve their lot".[37]

Forest management systems evolving since the nineteenth century colonial era were still largely premised on models propagating unilateral, centralized state control. The concept of social forestry was first highlighted in the silviculture conference of 1946. The uniform system was also experimented with, wherein plantations were raised under the canopy of a tree. India's 'tryst with destiny' began by accommodating the colonial structures with little 'reorientation' "in order to suffer in the interest of the country for the greater common good".[38] The creation of forest villages was perpetuated by the post-colonial government. Under the aegis of the independent government the rules became more stringent and the forest villages were reconstituted as forest jagir villages which involved the signing of a deed on part of the government and the villager or the plot holder. Pucca huts were constructed for the villagers in exchange for their services to the forest department.

As industrial and commercial interests overwhelmed the requirements of the defence and communication sectors the government and forest department sought new techniques of escalating the supply of timber. Social forestry and farm forestry, accentuated by the celebration of vanmahotsav in the 1950s, was successful in increasing the availability of construction poles, pulpwood and small timber. Social, community and farm forestry programmes were aimed at community participation in creation, protection and utilization of trees. They were designed to cater to the increasing availability of fuel, fodder and building material. They were also seen as effective means for generating employment and enhancing the revenue earnings of the forest departments. On a broader perspective they also had an educative purpose and were geared at stimulating general awareness on the indirect benefits of trees and forests in particular. The programmes also began to sensitise forest department staff to community needs.

Yet, the programmes were found lacking in a number of areas. They did not relieve pressure on natural forests, which continued to degrade. Communities were frequently dissatisfied with placing responsibility for plantation management with politicized panchayat organizations. In many communities, members felt that they would never receive any benefits from the project, and very often they did not. Finally, and most critically, social forestry programmes failed to respond to or resolve conflict between communities over rights to natural forest lands. This failure led to their continued overuse and had frustrated all attempts to stabilize forest use and allow ecological regeneration.

The word participatory forest management (PFM) first appeared in the forestry literature during the 1970s. The emergence of the notion of PFM can be traced from an experiment that was conducted in West Bengal during the early 1970s. Arabari, a small village in the Midnapore district of West Bengal where an Indian Forest Service official, Dr. Ajit K. Banerjee, started an experiment to protect the forests by involving local communities, ultimately gave the model of Joint Forest Management before the world. Banerjee, by profession a silviculturist was conducting an experiment of regenerating a sal forest at Arabari which was degraded due to indiscriminate felling of wood by the local people for their fuelwood requirements which culminated into uprooting of stumps of the Sal forest tantamount to direct violation of forest department rules. Banerjee tried to mobilize the villagers and sought their cooperation to protect the degraded forests at Arabari. It was under the style of a socio-economic project that generated alternative employment for the villagers in the lean season. A promise was also made to share 25 per cent of the benefit from sal timber after felling. This basic institutional reform motivated the villagers not to cut and sale trees that they had been practicing previously because of poverty and lack of employment. Hence during the first four years, the rapid growth of the sal coppice crops and its associations were noticed at Arabari.

The Arabari model proved that forests as a constant resource base for state revenue can be maintained by involving people in its protection and management. It improved the strained relationship between the Forest Department and fringe communities, resuscitated the sal forests, and enhanced the income of local people and even the Forest Department's revenues. With the implementation of the West Bengal Social Forestry Project of the 1980s, the idea of people's participation in protecting forests gained further acceptance. The Forest Department staff organized forest fringe communities into FPCs. These FPCs became formal institutions first in Southwest Bengal when the state government passed a resolution in 1989 specifying the composition, rights and responsibilities of FPCs. In view of the National Forest Policy of 1988, conservation of forests has been given priority with special emphasis on fuelwood and fodder development on available wastelands through people's participation to avert an ecological crisis and fuelwood and fodder famine.

Despite the initial euphoria JFM has found its vehement critics in NGOs, academicians, social activists. The overriding philosophy of the bureaucrats and most project implementers being, "we plan, you participate." Some scholars opine that, "the initial enthusiasm created by the JFM programme in West Bengal since 1988 and their overdependence on routine work coupled with "Jawahar Rozgar Yojana (JRY) fixation" have further deteriorated the prospects of a people oriented and indigenous natural resource generating community management strategy which the proponents of JFM programme envisaged in the early stages of its formation.[39]

Nandini Sundar has argued that the assumption upon which 'joint' forest management is predicated is of a historical dichotomy between civil society and the state, as if villagers and forest departments previously existed in water-tight compartments, locked into mutual antagonism. She further posits that this jointness is not new.[40] In fact creating spaces for co-management or community management has historically been as much a part of state management of resources as centralization. What is perhaps new is the degree to which it is being promoted by other actors or donors. Besides in the limited spheres of participation opened up by the state there is a joint construction of needs.

Since 1981 plants of different utility have been raised on road side land, canal banks, embankments, river banks, railway lands, institutional lands, private lands, forest department lands etc.[41] Forest Department launched several incentive schemes for local people to aid in protecting plantations. Actual work started with motivating the people in the area where land was available then the usual practice of raising the plants was initiated through the infrastructure available with the social forestry programme. Local people were encouraged to purchase the seedlings through Kisan Nursery under Guaranteed Buy Back

Scheme. Predicated on the status of the land, the protection and maintenance was handed over to the local bodies at different stages after identifying the beneficiaries through village panchayats. Hence committees had been constituted at various levels to exercise control beginning with the higher state echelons downwards for co-ordinating different government departments and target groups. These committees effectively controlled the activities to ensure the fulfillment of the professed objectives.

Panchayati Raj had initiated increasing democratization of these committees. PFM popularized the concept of Forest Protection Committees (FPCs) in Jalpaiguri. The message of Arabari experiment spread at a rapid pace within the forest department and foresters and forest department officials in other districts took initiative to regenerate the forests following the Arabari model. [42] Through informal discussion and group meetings, the circle conservator and his district forest officers intimated the field level foresters (beat officers and range officers) regarding the commitment to involve the communities in forest management. The field level officials in turn, started meeting the villagers and motivated them to protect the forests. Local party and Panchayat also played a pivotal role to involve the communities in forest protection. As a corollary some villagers came forward and formed their forest protection committees (FPCs). This enhanced the participation of local people in protecting and raising plantations.

After coming to power in 1977, the communist government initiated the decentralization of power by implementing three-tier Panchayat system at the level of Gram Panchayats, Panchayat Samities and Zila Parishads. The West Bengal Subordinate Forest Service Association was established in 1979 with a mission to mobilize local people and garner support for joint management. The formation of FPCs was further institutionalized by the government by introducing a series of legislations. [43] The government clearly spelt out the composition of the FPCs, its executive committee (EC) including its functioning and monitoring. The government of India ultimately passed the first order on JFM in June 1990 implementing the JFM in all its states and union territories. [44] The total number of FPCs was forty-nine, total area protected was 18,266 hectares and 12,169 FPC members were estimated in the year 1992.

Analysed in longue duree participation in forestry operations has been an age-old phenomenon. The Mughal period witnessed a pastoral population on the fringes of sedentary cultivation. They were actively involved in forestry operations and some groups even specialized in these ventures under the state apparatus. These communities were largely amalgamated within the expanding agrarian economy. [45] In case of economic regression their status could have been tenuous leading to social tension. [46] The process of transformation of forests from free good to fully commoditized good (in terms of economic theory) in the colonial period set up stresses and tensions which induced regulatory regimes for controlling access to and rights over forest space and resources. Such regimes might be regulated for and by estate holders wishing to assert their claims on the forest as their property, overriding and rejecting the claims of the forest dwellers invariably unleashing a period of conflict. The colonial government initiated a policy of conflict precipitating a discontinuity with its predecessor, the Mughal state. Eventually, the colonial government was forced to accept a policy of compromise for lack of indigenous knowledge in raising plantations and the crippling paucity of labour to work on their plantations. Compromise indicated a discontinuity in policy implementation with its predecessor tempered with certain qualitative differences, wherein village forests were created and villagers were forced to comply. The post-colonial period accommodated the policy of compromise but stringent policies and the high-handedness of forest officials had created pockets of anomaly in an age of democratization. Participatory forest management was a collaborative endeavour, though still in a nebulous stage. Increasing democratization of this institution could attribute to it nuances of success. Hence one can argue that Participatory Forest Management is part of a development discourse which is constantly being formulated and reformulated and is part of a historical process that has been evolving through the years.

Notes :

1. Chetan Singh, "Forests, Pastoralists and Agrarian Society" in David Arnold and Ramachandra Guha (eds.) *Nature, Culture, Imperialism - Essays on Environmental History of South Asia*, New Delhi, 1995, p. 30.
2. Chetan Singh, "Conformity and Conflict : Tribes and the 'Agrarian System' of Mughal India", in Muzaffar Alam and Sanjay Subrahmanyam (eds.) *The Mughal State 1526-1750*, New Delhi, 1998, p. 447.
3. *Ibid*, p. 447.
4. Neeladri Bhattacharya, "Pastoralists in a Colonial World", in David Arnold and Ramachandra Guha (eds.) *Nature, Culture, Imperialism Essays on Environmental History of South Asia*, *op.cit.*, p. 49.
5. Chetan Singh, "Conformity and Conflict : Tribes and the 'Agrarian System' of Mughal India", in Muzaffar Alam and Sanjay Subramaniyam (eds.) The Mughal State, *op.cit.*, p. 447.
6. Shirin Moosvi, *The Economy of the Mughal Empire, c. 1595 : A Statistical Study* (Delhi, 1987), p. 65. See also Harbans Mukhia, "Peasant Production and Medieval Indian Society", p. 245. As he mentions, "even a grand state of imperial dimensions was able to establish its overarching system in the region only with considerable degree of variation".

7. Chetan Singh, "Conformity and Conflict : Tribes and the 'Agrarian System' of Mughal India", in Muzaffar Alam and Sanjay Subramaniyam (eds.) The Mughal State, *op.cit.*, p. 425.
8. *Ibid.*, p. 425.
9. Gautam Bhadra, "Two Frontier Uprisings in Mughal India", in Muzaffar Alam and Sanjay Subrahmanyam (eds.) The Mughal State 1526-1750, *op.cit.*, pp. 474-90.
10. B.G.Karlsson, *Contested Belonging : An Indigenous People's Struggle for Forest and Identity in sub-Himalayan Bengal*, (Lund University, 1997), pp. 90-91.
11. *Ibid.*, p. 91.
12. Barun De, (et. al eds.), *West Bengal District Gazetteers, Jalpaiguri*, Calcutta, 1981, p. 58.
13. *Ibid.*, p. 54.
14. *Ibid.*, p. 55.
15. Chetan Singh, "Conformity and Conflict : Tribes and the Agrarian System" in Muzaffar Alam and Sanjay Subrahmanyam (eds.) The Mughal State 1526-1750, *op.cit.*, p. 429.
16. Chetan Singh, "Forests, Pastoralists and Agrarian Society in Mughal India", in David Arnold and Ramachandra Guha (eds.) Nature, Culture, Imperialism, *op.cit.*, p. 24.
17. *Ibid.*, p. 24.
18. *Ibid.*, p. 24.
19. Subhajyoti Ray, *Transformations on the Bengal Frontier- Jalpaiguri, 1765-1948*, RoutledgeCurzon, 2002, pp. 19-20.
20. *Ibid.*, p. 21.
21. *Ibid.*, p. 22.
22. T.N.Roy, "History of Forest Management in North Bengal", in *West Bengal Forest Commemoration Volume*, West Bengal Forest Directorate, Calcutta, 1964, p. 79.
23. Subhajyoti Ray, Transformations on the Bengal Frontier, Jalpaiguri, 1765-1949, *op.cit.*, pp. 21-22.
24. Chetan Singh, "Forests, Pastoralists and Agrarian Society", in David Arnold and Ramachandra Guha (eds.), Nature, Culture, Imperialism, *op.cit.*, p. 36.
25. *Ibid.*, p. 36.
26. *Ibid.*, p. 38.
27. *Ibid.*, p. 39.
28. Chetan Singh, "Conformity and Conflict : Tribes and the 'Agrarian System' of Mughal India", in Muzaffar Alam and Subrahmanyam (eds.) The Mughal State, 1526-1750, *op.cit.*, p. 429.

29. *Ibid.*, p. 429.
30. *Ibid.*, p. 430.
31. D.H.E.Sunder, *Survey and Settlement of the Western Duars in the District of Jalpaiguri : 1889-95*, Calcutta, 1895, p. 18.
32. *Ibid.*, p. 19.
33. *Ibid.*, p. 20.
34. Progress Report of Forest Administration for the Bengal Presidency for the year 1875-76, Calcutta, 1884, p. 38.
35. *Ibid.*, p. 39.
36. M.K.Raha, *The Rabhas of West Bengal*, Calcutta, 1967, p. 30.
37. *Progress Report of Forest Administration in the Lower Provinces of Bengal for the year 1968-69*, Calcutta, 1969, p. 10.
38. Jawaharlal Nehru speaking to villagers who were to be displaced by the Hirakud Dam, 1948, in *The Frontline*, January 1998, p. 10.
39. Nandini Sundar, "Unpacking the 'Joint' in Joint Forest Management" in M.Doornbar, Ashwani Saith and Ben White (eds.) *Nature, People, Power*, New Delhi, 2000, p. 35.
40. Chetan Singh, "Conformity and Conflict : Tribes and the 'Agrarian System' of Mughal India", in Muzaffar Alam and Sanjay Subrahmanyam (eds.) The Mughal State, 1526-1750, *op.cit.*, p.425.
41. I am grateful to Mr. Atanu Raha, Chief Conservator of Forests, West Bengal for his interesting inputs and ready guidance and to Mr. R.K.Chattopadhyay, Senior Forest Service, Burdwan Forest Division, West Bengal for the relevant data.
42. My interview with Dr. A.K. Banerjee was a rewarding experience as he shared the nitty-gritties of being a government official and the impediments to initiating a policy which later received international recognition. After four years of starting the experiment, Banerjee left the job in 1975-76 and joined the World Bank as a consultant. The message of Arabari experiment got international recognition due to the active interference of Banerjee publicising the experiment as one of the greatest successes of the forest department through several publications and World Bank Policy papers.
43. Report of the Forest Administration of Bengal, 1977-78, Calcutta, 1978, p. 15.
44. Government Forest Act, MoEF, No. 6.21/89-F.P., Implementing JFM in all states and union territories.
45. *Ibid.*, p. 428.
46. *Ibid.*, p. 429.

Communication in North-East India During Pre-Colonial and Post Colonial Days

W. Jyotirmoy Singh

India's Northeast spreads over 2,55,000 sq.Km with a relative population of over 40 million. It consists of the states of Tripura, Mizoram, Manipur, Nagaland, Arunachal Pradesh, Assam, Meghalaya and Sikkim. It comprises about 8 per cent of the land area and 3 percent of the population of the country.

The North East Region shares around 4849 kilometres, that is 98 per cent border with her neighbours—Nepal, China, Bhutan, Bangladesh and Myanmar. It means that the North East region shares only 2 per cent land contiguity with the rest of India through a 22 km wide Siliguri corridor often referred to as the 'chicken's neck corridor.

This paper will briefly delve on the communications of the Northeast region of India down the ages. By communication, here it means the mode of journey undertaken by one from one place of north east India to another part and beyond. Thus, this paper deal on I) communication between the north east India and its neighbour in the pre-colonial and colonial day, II) communication between the north east India and its neighbours in the post independence period and III) what can to be done to improve the communication in the north east region of India.

Communication in Northeast India in pre-Colonial and Colonial period:

As the North East region is far away from national markets, it had a natural trading advantage with the neighbouring countries. Cross-border exchanges form an important parameter in Northeast region's relation with its neighbours. People living in North Eastern Region carried out both internal and external trade with their immediate neighbours like Myanmar, Tibet, Bhutan, etc. This sustained the socio-economic polity of the people of the region.

Since the early days Assam carried on trade with Bengal. There existed four routes from Assam to Bengal, one by water and three by land. The river route starts from Goalpara. It runs down the Brahmaputra and via the Jennai from Jamalpur and then after some distance along the Prabha river, a navigable branch of the Ganges. The journey by boat used to take 25 to 35 days to complete the voyage from Goalpara to Calcutta, and 33 to 43 days from Calcutta.[1] The first land route from Bengal to Assam lay through Murshidabad, Maldah, Dinajpore, Rungpore, Bagwa and Goalpara. This routewas almost impossible during rains. The second route was via Dacca, Dumary, Pucuole, Jamalpore, Singimari and Goalpara. This route was also impossible in the rains. The third route passed through Sylhet, Cherra, Moplung, Nungglao, Ranigaon, Khanamukh and Gauhati.[2]

Assam also had trade relationship with Tibet and Bhutan. M'Cosh refers to as many as five routes leading from Sadiya into Tibet or China proper.[3] Trade with Tibet and Bhutan seems to be more profitable to the traders of Assam.

In the pre-independence days Barak-Surma valley was an important centre in the region for inland and overland trade. The eastern portion of the valley constitutes the present day districts of Cachar and Hailakhandi of

Assam.They formed a part of Tripura kingdom and Dimasa kingdom respectively during the medieval period. The Western portion covered the major part of the valley.It constitutes modern Karimganj district of the state of present Assam and Sylhet division of modern Bangladesh. It was commonly known as Sylhet (Lower Valley) during the medieval and British period. [4] The proximity to Chittagong has contributed to Sylhet's commercial prosperity since early times. In the medieval period, Sylhet was the gateway for the neighbouring areas for trade with Bengal. Thus, Sylhet was an important trade centre for the Khasi-Jaintia hills, Cachar, Manipur, Mizo (Lushai hills) and Tripura.

Sylhet was separated from the south of Assam Valley by the Garo Hills, the Khasi Hills and the Jaintia Hills. Communication between Sylhet and Assam in the past was not an easy one. On the other hand, there was no natural boundary between Sylhet, Cachar and the neighbouring districts of East Bengal and communication was unhindered.[5]

Sylhet carried out trade relation with Khasi Hills, Jaintia Hills and the Garo Hills. Infact, the entire volume of the flourishing Khasi-Jaintia trade with Bengal passed through Sylhet. [6] Lime along with iron ore of the Khasi Hills were in considerable demand in Sylhet from where it was transported to Kolkata and other parts of the country. Lime trade of the Khasihills was a profitable one. [7] The Garo Hills, on the other hand, supplied to Sylhet cotton, agar, lac and other forest product.

During their colonial hegemony, the British introduced a number of measures, both administrative and otherwise to develop the areas. One such measure undertaken was the construction of road, linking one area to another. W. J. Allen in The Report of 1858, recommends the annual repairing of the old road between Sylhet and Assam passing though the hills, viz. Jaintiapur and Jowai to Jaggee Chokey in

Nowgong. This was a very useful road to the people of the Jaintia Hills. It was also their route of direct communication with Sylhet and Nowgong. A considerable traffic passed along this road to Sylhet and Nowgong.

Since 1933, there has been a well maintained motorable road which connects Shillong with Sylhet through Dawki, one of the market centres in the Jaintia Hills. The same road connects Shillong with Gauhati. This road i.e. Sylhet to Guwahati through Shillong became the lifeline of trade and commerce. It also played an important role because it connected all the major and important and administrative centres, both in the Surma and the Brahmaputra valleys. [8] Thus, a good motorable road existed between Sylhet to Guwahati through Shillong in the pre-independence days.

Sylhet also carried out trade relationship with Tripura since early days. The Cacharis carried out trade with Sylhet via Mulagal.During the reign of Yashonarayan (1583-1601),they mainly marketed rice from Sylhet and the neighbouring areas. [9] There are also records of trade route from Sylhet which went to China through Assam, Manipur and North Myanmar. The traders from Kamarupa, Manipur, China, Myanmar and other neighbouring areas traded with the valley. Sylhet, thus acted as the nerve centre of the communication between the North eastern region and the region beyond it.

The minerals and the forest resources have been the focus of attention of the British. The NER (details?) have been producing materials like tea, jute, cotton, mustard, cotton oil and coal and timber. To support the production and transportation as desired by the colonial power, motor roads, navigation and rail road have been developed in a coordinated manner.

The growth of the tea industry in the Northeast in the 1830s was a landmark in the development of transport and communication in the region. The Assam Tea Company plied a steam boat from Gauhati to Calcutta in 1842. The Indian Steam Navigation Company provided two vessels in 1860 to ply between Calcutta and other parts of Assam valley at an interval of every six weeks. [10] In 1874, a daily mail steamer service was introduced by the Joint Steamer Company (The Indian General Navigation Company and Railway Company Limited and the Rivers Steam navigation Company Ltd). In 1884, steamers started plying daily between Dibugarh and Dhubri. A steam service from Calcutta to Dibugarh took 4 ½ days and that took 3 ½ days for the return journey .

By the end of the nineteenth century, tea plantations were extended to different interior areas far away from the centre of shipment This necessitated another mode of transportation suited according to the needs. In 1880, the Assam Railways and Trading Company, was formed. It opened the Dibru Sadiya Railway in a couple of years. It linked the tea gardens along the Dibugarh steamer ghats and Jaipur Road. About the same time, the Company opened the coal mines at Makum and Ledo and extended the railways up there. It thus facilitated the transportation of coal, both to the tea gardens and to the industrial centres at Calcutta along this route. The success of the Company encouraged the opening of railway in different part of Assam. In 1885, the Government opened the Jorhat Provincial Railways and a decade later, the Tezpur Baliapara Railways. [11]

Chittagong was joined by railways with Dibugarh in 1904and , by that time all the centres of plantations were linked to one another by railways. In 1910, Amingaon was linked with Calcutta by the railway line called Assam Bengal Railways. This greatly reduced the cost of transportation of produce from Assam and also facilitated the easy movement of people from one place to another. [12] Thus, by 1947, because of tea, rail network gradually connected the north east, especially Assam with Bengal.

Tripura also occupies an important geographical location in the region. Located partly at the eastern extension of the Bengal plains it was in the immediate vicinity of the hill region to its east. [13] It had trade relation with Sylhet and other adjoining districts of Bengal since the early days. The rulers of Bengal looked upon Tripura as a rich source for supply of elephants. The forest products timber, bamboo and cane were supplied to Bengal in exchange of consumable items. Timber trade also existed between Tripura and Sylhet as early as the 11th century AD. [14] The Report of the Administration of Bengal, 1888-89 records that, "The trade was in the hand of Bengalis. They export cotton, oilseeds and forest produce of various kinds, and imports all necessaries of life (except rice) such as salt, dried fish, pulses, piece goods, hardware and pottery. The export trade is chiefly maintained on the banks of the rivers and the import trade is naturally conducted on the border of the British districts. [15]

The British Expansionist policy, which was followed by the expansion of trade and commerce in India, also had its effect on Tripura. Railways were built in Bengal with the object to supply to Calcutta raw materials and carry the manufactured goods from there to the consumers in East Bengal and other areas. As already mentioned the Assam-Bengal Railway was constructed in the later part of the 19th century. The Assam-Bengal Railway connected the sea port of Chittagong to Silchar. Rail lines along the frontiers of Tripura were construced. Akhaura was the nearest railway station to Agartala. A metalled road linked Agartala with Akhaura. The opening of the frontiers like Akhaura, Khowai, Kamlasagar, Comilla and Belonia helped in the mobility of men and material. Thus, one can perceive that before India's independence Tripura was well connected.

Manipur lies on the route between South Asia and Southeast Asia. Inspite of the difficult and hilly terrain, trade routes connecting Manipur with her neighbours were known to have existed as early as the second century B.C. [16] There were routes through the hills connecting Manipur with the Surma and the Brahmaputra valley in the west and the Yunan Province of China and Myanmar in the east.

Writing in 1835, Pemberton observed that merchants from Bengal used to have ample time to meet the Burmese and the Shans from banks of the Irawadee and Ningthi (Chindwin) rivers, affecting the sales of their goods and return to their respective countries before the commencement of the rainy season, when trade was necessarily suspended. [17] Cheitharol Ital Kumbaba, the *royal chronicle* of Manipur also records a long distance trade between Manipur and her neighbours like Tripura, Assam, Myanmar and Yunan Province of China.

Pemberton also says that even in the early 19th century, a very brisk trade was carried with the inhabitants of the Kabow Valley of Myanmar and the merchants of the southern Chinese province of Yunan with Manipur. [18]

Cheitharol Kumbaba, also records the trade between Manipur with Yunan province of South China in 1630. From the Chinese, the Manipuri learnt the art of manufacturing gunpowder.[19] Among the items of trade carried out between the two were horses, ivory, wax, clothes and ponies.[20]

Regarding the nature of the traffic with Mayanmar, Pemberton states that from the middle of December to the latter part of April, the towns of Bhumo (in north Myanmar) presented the most animated scene of active industry and greater variety of tribes perhaps than in any other fair in Asia. From every part of the Burmese empire, merchants used to come together at Bhumo at that time of the year and after having effected their purchases, traversed the country in every direction by land and water routes, for the purpose of again disposing of their goods. The people of Manipur frequently journeyed for this object to the border of Yunan, into the lower Shan country occupying the immediate tract between the river Ningthi and Bhumo.[21]

In the 16th century, there was an established commercial contact between Manipur and Assam. Trade with Upper Assam started in 1536. Jai Singh (1763-1798) is said to have gone to Tekhao (Assam) through Maram (in northern Manipur) and Thibomei (Kohima). Among the other developmental works that Jai Singh undertook was the improvement of the road to Assam, which benefitted the trade and commerce of the people of the hills residing along this route.

There was a brisk trade between Sylhet and Manipur, and Cachar played the role of a buffer in these transactions. The Raja of Cachar collected duty on boats coming from and going to Manipur in the river ghats at Sealtekh and Jirighat.[22]

Trade was mostly confined to items like wax, ivory, tusk and livestock like buffaloes, ponies and elephants which were also major items of trade. Cattle constituted an important item of export. Major portion of the cattle exported from Manipur were imported from Myanmar.[23]

There were three important routes that connected Manipur with the Kabow Valley of Myanmar. They were Heirok, Aimol and Marringroutes. The traders undertook the route according to their convenience and commercial interests. Important western routes that connected Manipur with Cachar and beyond were the Aquee route, the Kala Naga route and the Khongjai route. The point of departure for these and every other well known route from Cachar to Manipur was Banskhandee, a village near the eastern extremity of the cleared plains of Cachar. [24] Thus, Manipur had well known route to Cachar, Tripura, Sylhet and Dhaka. [25]

The king of Manipur placed high priority on the construction of roads- both internal and external road. [26] According to the Treaty of Badarpur (1824), Govinda Chandra, the king of Cachar was to pay an annual tribute of Rs. 10,000 annually. In 1827, the British allowed an exemption from the payment of revenue for the two preceding years, in return the Raja was required to construct a road from Sylhet border to Manipur through Cachar.[27] The British officers lauded the king of Manipur for the successful construction of the road from Manipur to Cachar.

To sum up, one can understand that in pre- independence time, India's Northeast region had communication infrastructure within its region and also beyond it. There also existed road, riverine and rail connections with their neighbours. People and also the materials of the region moved from one place to another and beyond conveniently and also transacted economic activities.

Northeast India After1947 :

India's independence and subsequent partition in 1947 was associated with , the displacement of a large section of people in the Indian sub-continent. However, one often overlooks the difficulties created to the North East India by the partition of India. After India's independence in 1947 and its subsequent partition in its eastern sector, the north east was transformed into an enclosed landlocked landmass.

The newly enclosed landmass, i.e. North Eastern Region, is connected with the Indian heartland through a slender 22 kilometre Siliguri corridor, which now became the Gateway to the Northeast. No more than two per cent of its land is contiguous with the rest of India. The remaining 98 per cent is bounded by international boundaries.

The border conflict with China in 1962 and the deterioration of India's relations with Pakistan in the mid 1960s led to a disruption of rail, road and river links between the North East and other eastern states and the neighbouring countries. Road and railway's transit routes through erstwhile East Bengal were lost, as the region now came to known as East Pakistan after India's Independence. [28] India also lost its access to the Chittagong Port. It crippledNorth East's economic linkages and driven up costs of all commodities. [29]

Prior to India's independence and its subsequent partition, one could travel from Agartala, the capital of Tripura which lies in the southern part of the Northeast India, to Kolkata within a matter of day only. The total distance between the two was around 350 Km. After independence, Tripura is surrounded by Bangladesh to the north, south and west. Its international border is 856 km long (84 per cent of its total border-length). The State is connected with the rest of India by only one road NH-44, which runs through the hills to the border of Cachar District in Assam. Now, if one wants to travel from Agartala to Kolkata by road, one has to go by the one and only Siliguri Chicken Neck Corridor travelling a distance of 1,700 Kilometre, taking a journey of more than three days.

Till India's Independence, Karimganj was an important trade centre of Assam province after Guwahati and Tinsukia. Its commercial importance suffered a setback due to disruption of communication link with the rest of India. Similarly, after independence, other North Eastern Regions like Cachar, Khasi-Jaintia Hills and Garo Hills lost their easy access of trade in commodities, specially with Sylhet. Thus, the relation of North East India with its neighbours like modern day Bangladesh and Myanmar are sealed because of its enclosement following India's independence and its subsequent partition.

After independence North East lay far away from national markets. All products consumed in the North East comes from distant regions in India. India spends huge amount in transporting food items like pulses, poultry products, grains etcto NER for which the people have to pay more to meet the cost of transportation. Likewise, prices of all other items that comes from our hinterland are higher in the region reducing the affordability of its people.

To compound the matter, communication especially in surface transportation in Northeast India after India's independence is not encouraging. One, however, needs to note that the Sixth Sectoral Summit of NEC on Rail Connectivity Programme on 19th June 2007 states that the total rail network in the northeast is only 2578 k.m. which is 4% of the total rail network of the country. One of the main reasons for the low connectivity of the railway link in the north east can also be attributed to the partition of India. Partition seriously affected the rail connectivity in the North east.

The Assam Railway Company which was formed in the late 80s of the nineteenth century, split up according to the political boundaries and the portion of the system at the time of partition. About 2,603.92 km long railway line of the Eastern Bengal Railway fell into the portion of East Pakistan (renamed as Pakistan Railway in 1961). The Partition led to the complete severance of Assam, especially after the Indo-Pakistan Standstill Agreements came to an end on 29 February 1948. It disrupted normal channel of communication with the rest of India. It was restored with the completion of the Assam rail Link Project with its Headquarters at Pandu in January 1950.

In the pre-partition days rail travels from Gauhati to Kolkata took about 24 hours, with one ferry crossing. After partition, the journey used to takes about 42 hours with two ferry crossings. However, after the completion of the two bridges at Pandu over the Brahmaputra in 1967 and Farakka Barrage over the Ganga in 1971 and the conversion of the entire line up to Guwahati to broad gauge, it has come down to pre-partition duration. [30] Nevertheless, till the late 1960s, if one was travelling from Gauhati onwards, one had to change train at Barauni for embarking on a broad gauge. This change of train for travelling by broad gauge continued till late 1960s at Bongaigaon and by the early 1980s at Gauhati.

When the rail connectivity of the North east is very low and its network connection is carried out at a minimal pace, it is easy to understand the fate of the pace of development in the north east and its resultant effect on the price of the commodities imported from other parts of India. This situation still prevails in different parts of the North east.

Remedial Measures to rectify the difficulties of the North eastern Region:

As already stated, the **Northeast Region's** traditional communication links were disrupted by the partition of the country.

The northeastern states could easily procure many of the goods required by it from across the border at a fraction of the cost of goods transported from distant Indian destinations. Countries like Myanmar and Thailand - next door to Northeastern region can provide many of the items at much low price because of cheap in the cost of carriage. Even Bangladesh can supplement our requirement of fish, which is a staple item. Here in this regard it is necessary to point out that in almost every meal in a Manipuri household, a meal of fermented dry fish is an essential item. Nowadays, dry fish worth crores of rupees are annually imported from outside Manipur from places like Assam, Gujarat and even Bangladesh for fermenting. Dry fish imported from Bangladesh, through overt and covert transaction, is the most preferred one because of its quality and suitability for fermentation. An Indo-Bangladesh treaty on transaction of dry fish will be beneficial to both sides. The legalisation of this trade will generate revenue to Bangladesh and in Manipur it will reduce the cost of procuring dry fish.

Efforts should also be made to have a transport corridor from Agartala to Kolkata via Dacca or from Sylhet to different parts of North East India. This will tremendously help in reducing the time of travelling from the southern portion of the Northeast , like from Tripura and Mizoram to Kolkata. Time taken in transportation of goods will also be reduced, thereby reducing the price of the commodities. Some years back Kolkata to Dacca bus service was started. One needs to ask why it took so long to make Agartala to Kolkata bus service through Dacca a reality.

The opening up of the frontier between the two countries will also be beneficial to both the countries. Scarcity of mineral resources, except natural gas, has been a major cause of hindrance in the development of Bangladesh. The Northeast Region with its mineral resources, agricultural and forest resources, hydro-power potentials, can be of help to the development of Bangladesh. Only 2.25 of India's products are exported to Bangladesh. In this sense, it is not surprising to note that till 1997-98, Bangladesh ranked just 11th in the list of India's overall export.

For long, Myanmar had been ignored. It is too important a country to be ignored for its trade and security considerations. Northeastern States of India altogether share 1643 km Indo-Myanmar border. Arunachal Pradesh, Mizoram, Manipur and Nagaland share 520 kms, 510 kms, 398 kms and 215 kms respectively with Myanmar borders.

Myanmar is the gateway to India's 'Look East' policy. Since the beginning of the 90s of the last century, India is going all out to strengthen its relationship with Myanmar to achieve its stated objective. As a result, the two countries signed a trade agreement on 21st January 1994. The trade agreement became operationalised on April 121995 through Moreh, the border of Manipur and with Tamu the border town of Myanmar. Subsequently, 22 items have been allowed to be exported and imported under the free trade agreement signed between India and Myanmar. They include *mustard seeds, pulses and beans, fresh vegetables, fruits and soyabean imported from Myanmar*. On the other hand, *India supplyclothes, shoes, medicines, woolens and engineering goods* to Myanmar. These items are in great demand in Myanmar.

Permission is granted for one to travel from Moreh in India to Kalewa in Myanmar. There is however, a demand that the Indian government should introduce International bus service linking Moreh with Mandalay. The bus service linking Moreh with Mandalay would not only help the traders, but will also bring tourists from Myanmar to India.

Development of road transport will also promote travel and tourism between the two neighbours. The opening of land route could also make possible for India in reaching ASEAN countries through land because Myanmar is the only ASEAN country having a land border with India.

One notable incident broke out in Manipur in the early part of this decade. Blockade was imposed in Manipur on National Highways 39 and 53 for fifty eight (58) days in 2005. The two National Highways are lifelines of Manipur, supplying from the rest of India the commodities consumed in Manipur. Fortunately, the people of Manipur offset this difficulty by procuring goods like food items and petroleum products from Myanmar through Moreh, its border town. In 2011 also, various non-governmental organisation gave calls for economic blockade for 68 days. Petroleum products and other goods were imported from Myanmar to meet the shortage of Manipur. In view of the difficulties to transport rice from outside Manipur through NH 39 and 53 respectively, the Government of India had gave permission to the Government of Manipur to import rice from Myanmar.

In view of India's need for developing economic relation with its immediate neighbours like Bangladesh and Myanmar, their arise the need to expand India's economic link outlook with its neighbours and beyond. People of the North Eastern Region and the state Governments of the region are calling for the re-opening of the old trade route. India has made road and rail project to link the North east with Myanmar. With a view to realising the Trans Asian Highway through the North east, India has already helped Myanmar by constructing road within that country to a good extent. At present the Indian Railway on the line of Trans Asian Railway is carrying on its task of linking Moreh, the border town in Manipur with Imphal by train.

There is an old saying in Manipur that the "Eastern Gate" of Manipur that is closed will be re-opened after sometime. Here, the "Eastern Gate" means Manipur's eastern neighbour i.e. Myanmar. It means that Manipur will resume her relation with her eastern neighbours like Myanmar. India is pursuance of 'Look East Policy' subsequent to the signing of Indo-Myanmar trade agreement in 1994 has been taken by some section in Manipur as the realisation of the old saying i.e. the re-opening of the "Eastern Gate" of Manipur.

Notes :

1. After proceeding up the Prabha river for two or three days the boat used to reach to the Ganges which then ascended for three or four days more till it reached to the mouth of the Matabhnaga or the Jelligi down either of which they proceeded to Calcutta. As these two rivers almost became dry during winter boats followed the route through the Sunderbans.
2. H.K. Barpujari, *Assam in the Days of the Company*, NEHU, Shillong, 1996, p. 276.
3. They are: the pass of Dibong, the Mishmi route, the Phugan pass to Manchee and China, the route by Manipur to the Irrawady and the Patkai pass to Bhamo on the Irrawady. The most important and easy route was on the north- eastern side over the Patkai to the upper districts of Myanmar and thence to China. Through this route Shan invaders came to the Brahmaputra valley. In 1816 during the Burmese invasion, some 6,000 Burmese troops and 8,000 auxilliaries crossed the Patkai into Assam. In the former days, the Burmese government took care that there should be a village, or rather a settlement every twelve or fifteen miles along the route and it was the business of the peole to cut the jungles and remove all other obstructions from the path. S. L. Baruah ,*A Comprehensive History of Assam*, Delhi, 2003, p. 439.
4. M.S.N. Rahaman, Trade and Commercial Activities of medieval Barak-Surma Valley, Proceedings of *NEIHA* 28th Session, Goalpara, 2007, p. 222.
5. *Ibid*, p. 222.
6. J.B. Bhattacharjee, *Trade and Colony*, NEIHA, Shillong, 2000, p. 11.
7. This can be ascertained from the fact that in 1778, the District Collector of Sylhet, Robert Lindsay drove away all the Europeans Greek, Dutch, French, Portuguese and Armenians who were engaged in lime trade and engaged himself in the trade. By the time he left Sylhet for England, he made a fortune from the lime trade.
8. L. S.Gassah, *Trade Route and Trade relation between Jaintia Hills and Sylhet District in Pre-Independence Period*, Proceedings of *NEIHA* 9th Session, Gauhati, 1988, p. 488.
9. M. S. NRahaman,op. cit. ,p. 224.
10. S. L. Baruah, op. cit., Delhi, 2003, p. 644.
11. Ibid, p. 644.
12. Ibid, p. 645.

13. SukhenduDebbarma, Trade Relations Between the princely states of Tripura and Bengal in the 19th and 20th century, Proceedings of *NEIHA* 22nd Session, Tezpur, 2001, p. 288.
14. J.B. Bhattacharjee, op. cit. , p. 12.
15. SukhenduDebbarma, p. 289.
16. G. Ladhya, *Early Indian Economics*, Bombay, 1966, p. 133.
17. R.B. Pemberton, *Report on the Eastern Frontier of British India*, 1966, pp. 174.
18. *Ibid*, pp. 34-35.
19. T. Chodson, The Meitheis, Delhi, 1989, p. 21.
20. During the reign of Garibniwaz (1709-1748) horses from China were brought and perhaps brought from the Chinese merchants by the kings. During the early 19th century, Chinese merchants from the Yunan Province also visited Manipur, carrying away in return ivory, wax, clothes and ponies.
21. R. B. Pemberton,*op.cit.*, p. 32.
22. J.B. Bhattacharjee,op.cit., p 12.
23. In 190708, out of the total cattle 5753send from Manipur to Cachar and Golaghat in Assam, 3,240 cattle were imported Myanmar. In 192627 out of 3,636 cattle exportedto Assam, 2799 buffaloes and 109 ponies were imported from Burma. This trade was mainly done by the Mohammedans from Surma valley.
24. The Aquee route was the northern most route. Its total length from Banskhandi to Jainagar (also known as Sangaithen) in Manipur was about 87 miles of which the first 30 1\2 miles passed through a dense forest, intersected by numerous streams. This route crossed five ranges of hills. The Kala Naga route from Banskhandi to Lamangdong (Bishenpur) was also known as the TongjeiMaril by the Manipuri. Its total distance was 82 ½ miles. This route went across eight ranges of hills. The advantage of this route was that one could cross the Barak at a point in Jiri, on a boat carrying around 500 maunds of load, even in February, when the water in most of the rivers was generally low. The third route was called the Khongjai route. This route principally passed through the areas inhabited by the Khongjais. The villages on this route were few and small. The elevation on this route was less than that of the Aquee and the Kalanaga route. Above these routes, there was another route to Tekhao (Assam) which went by the north through Thibomei (Kohima) and Maram.

`25. To get a fair idea of the route from Manipur to Assam and further west, it would be worthwhile to follow the route passed by Jai

Singh (1761-1798) on his pilgrimage to Nabadwip in 1798. Jai Singh on his way to Cachar passed through Bishenpur (which is 17 miles south of Imphal). Then going by the course of the Leimatak River, he passed the village of the Chirus and the hill ranges of the Ruyang. Then he proceeded by the course of the river Irang, a tributary of the Barak River, thereby reaching Cachar. From, there he further proceeded by the course of the Barak River to Bhanga, Surma and Sylhet. He then went to DakhaDakshin. From there he proceeded to Tripura and along the course of the rivers Meghna, Padma and Brahmaputra and reached Nabadwip in Bengal.

26. In the 18th century, Garibniwaz deputed a *Katwan* (an official) to construct a road for MayangLeima, a Manipuri princess who was married to a Prince of Cachar (Mayang).CheitharolKumbaba recorded that Jai Singh constructed a road upto the river Gwai (Barak) i.e. uptoCachar. This road was further improved by Chourjit (1803-1813) in the first decade of the nineteenth century. In 1832, Gambheer Singh (1826-1834) constructed a road upto the Brahmaputra valley via the Naga Hills. *Singh.L. Ibungohal&Singh N. Khelchandra, CheitharolKumbaba,* Imphal, 1989, pp. 111, 135.
27. J.B. Bhattacharjee, op.cit., p. 33.
28. Chakravati, Mahadev, Social Change in North-East India, *NEIHA*, Shillong, 2008, p. 23.
29. *Ibid*, p. 16.
30. *Ibid*, p. 23

AMBIGUITIES OF CULTURE

The Book and the Civility Paradism : A Study in the Dissemination and Continuation of the Shakespearean Tradition in Colonial and Post 1947 India

Kaushik Chakraborty

The colonial rule in India started from 1757 when the English East India Company defeated the Bengal Nawab Siraj-ud daula in the battle of *Palashi*. After the initial experimental years the colonial rule started to get a consolidated form from 1793 onwards with the introduction of the Permanent Land Revenue Settlement in that year. Introduction of printed books in India in its modern form started under the colonial rule of India. Reading was present in pre-colonial India also. The most widely practiced form in pre-colonial India was the oral tradition which was mainly prevalent in rural areas where *Ramayana, Mahabharata, Panchalis* and various other secular and religious texts were read by a reader or a *Kathak* surrounding him the listeners. In pre-colonial period there existed already a huge amount of vernacular, Sanskrit and Persian works. This literature, indigenous in character, had a distinct *mode* of its own. Under colonial rule starting from the last quarter of 18th century down to mid 19th century print culture was introduced and grew up in Bengal as well as in India. With the process of the consolidation of the colonial rule the perception of the colonial rule started to differ among the different classes of the indigenous society. In 1831 when TituMeer, in a stone throwing distance from Calcutta was organizing a revolt and was proclaiming the end of the colonial Raj, at the same time in his letter to Victor Jacquemont Rammmohan Roy wrote that India needs many more years of British rule. [1] It is necessary to locate the origins of such purely opposite perceptions. One significant thing is that men like Rammohan Roy were among the first lot who were being introduced with printed books under colonial rule. The present study aims to locate how the colonial rule was perceived by different classes under colonial rule. The area of this study is modern Bangladesh, West Bengal, Bihar and Jharkhand.

The colonial rule brought with it the civilizing agenda. The biggest organ which propagated this civilizing agenda was the *book*. The general argument of the civilizing agenda was that the colonized subject people required to be elevated to a state of life and culture which can be termed as a civilized life and as a whole a civilized society. This implies that the colonial state perceived the indigenous society and culture of India as something which can not be termed as civilized. So the colonial state perceived the society and culture of pre-colonial India as sub-civilised or uncivilized and it was the printed book which helped in the dissemination of this perception. This automatically brings the question *was the pre-colonial state and society a despotic or uncivilized one or does there existed a state and society in pre-colonial India with its own civility paradigm? If there existed a state and society in pre-colonial India with its own civility paradigm then why the colonial rulers perceived it as sub-civilised or uncivilized? Does this mean that the colonial state applied a distinctly different civility paradigm to characterize and understand India? If so, then why the colonial rulers thought in this way?*

In Europe from Reformation to Enlightenment and in the post Enlightenment period the growth of print culture accompanied by books played a crucial role in the shaping up of the modern European society. In Europe print culture originated from within the society and not came from outside. But in India since the book was introduced under the colonial rule and hence not from within but from above the society so with print culture came new form, content and above all a distinct new *mode*. The new *mode*, firmly embedded in Shakespearean tradition gradually disseminated Shakespearean tradition which entrenched in the indigenous literature also. This *mode* guided the mode of literary tradition in India during the entire colonial period. So what is necessary to locate is, *was there a continuity of literary mode between the pre-colonial and the colonial period or was it a change? If it was a change, then, what was the nature and character of the change? If it was a change, then does it affected reading and interpretation? Does interpretation and reading habit resulted in the growth of a new vision about state and society?* Finally, *what is the nature and character of the post 1947 book? Is it playing a liberating role?*

The Book and civility paradigm in pre-colonial India

By the term book in pre-colonial India it appears that it will be correct to say that in pre-colonial India books existed in manuscript form. Direct reading of book in ancient times, as it appears was limited among the learned circle because most of the books in ancient times were written in Pali or Sanskrit. For example, *Arthashastra* of Kautilya, *Sakuntala* of Kalidasa or *Harshacharita* of Banabhatta. Regional literatures developed much later which were vernacular like, *Charyapada* in Bengal. During the Sultanate and Mughal period on the one hand Persian literature flourished, like, *Tabakat-I-Nasiri* of Minhaju-s-Siraj, *Ain-I-Akbari* of Abul Fazl and several others. At the same time flourished regional literatures like Bengali literature under Hussain Shahi Bengal, for example, *Paragali Mahabharat, Vaishnava literature,* although many in Sanskrit, *Mangal Kavyas,* the *Pir literature* etc.

In pre-colonial India, it appears that there existed a distinct conception of civility. There is no doubt that in pre-colonial India the manuscript book never acquired a large or wide number of circulation. Rather, the book in pre-colonial India generally remained confined within a closed circle of learned priests, pundits, moulvies, Kings or Sultans and their learned circle. It was the oral tradition which actually helped in spreading the content of the book among a wider section. This indicates that the book through oral tradition disseminated ideas and perceptions among a wider section. The pre-colonial manuscript book is a source to locate this paradigm of civility. Instead of practicing despotism, it appears that a concern for the civil society always remained as the agenda of the pre-colonial state. In his *Arthasastra* Kautilya has observed that the king who is well educated and disciplined in sciences, devoted to good government of his subjects, and bent on doing good to all people will enjoy the earth unopposed. [2] This idea of well being of the subjects continued in the Sultanate and Mughal period also. In his *Tarikh-I-Firuz Shahi* Zia-ud-din Barani has observed that Sultan Ghiyas-ud-din Tughlaq always cared about the cultivators, Sultan's understanding was that revenue had to be collected in a way so that raiyats should increase their cultivation and there should not be such exaction so that cultivation fall off and no increase happen in future. [3] In his autobiography *Futuhat-I-Firuz Shahi* Sultan Firuz Shah Tughlaq observed that he became the king because it was god's desire and blessings. [4] Sultan observed that due to god's favour his heart used to be occupied with an earnest desire to do well being of the poor and needy and so he always tried to suppress wickedness, irreligion and tried to establish the true. Sultan also narrated that he punished both the Muslims, Hindus and a Muslim religious leader who claimed himself as Prophet. Sultan Firuz Shah Tughlaq narrated that when one Ahmed Bahari claimed him a prophet the Sultan stopped the practice of that sect. [5] Again the same Sultan banished all frivolous taxes and collected only *Kharaj, Jakat, Jizya* and *Khams*. In his *Tarikh I Firuz Shahi* Shams-I-Siraj Afif observed that Sultan Firuz Shah Tughlaq constructed two canals, *Rajiva* and *Alaghkhani* which resulted in advancement and prosperity of agriculture in the Hissar-Firoz Shah region. [6]

Abbas Khan Sarwani in his *Tarikh I Sher Shahi* specially noted that Sher Shah was not only concerned about the cultivators but was also caring. During military expeditions if unavoidably crops were destroyed Sher Shah used to send amirs with surveyors to measure the amount of the cultivated crop destroyed and he used to pay compensation then and there. [7] In his *Ain-I-Akbari's* preface Abul Fazl observed that a true king implies a king whose object becomes removal of oppression and providing security, health, justice etc [8] and this was his praise about Mughal emperor Akbar. Abul Fazl narrated that silly and shortsighted men can not distinguish a true king from a selfish ruler. Nor is this remarkable, as both have in common a large treasury, a numerous army, clever servants, obedient subjects, an abundance of wise men, a multitude of skilful workmen, and superfluity of means of enjoyment. But men of deeper insight remark a difference. In the case of the former, the things just now enumerated, are lasting; but in that of the latter, of short duration. The former does not attach himself to these things, as his object is to remove oppression and provide for everything which is good. Security, health, chastity, justice, polite manners, faithfulness, truth, an increase of sincerity etc., are the result. The latter is kept in bonds by the external forms of royal power, by vanity, the slavishness of men, and the desire of enjoyment; hence, everywhere there is insecurity, unsettledness, strife, oppression, faithlessness, robbery. Not only that. Abul Fazl also emphasized that a king should not be rigid in mind and should be intelligent and open minded. In Abul Fazl's own words, "In his wisdom, the king will understand the spirit of the age, and shape his plans accordingly".[9]

In his *Akbar Nama* Abul Fazl has observed that when Emperor Akbar learnt about *Satidaha* system (burning of Hindu widows) he appointed inspectors in every city and district to see and to prevent if it was an incident of forceful burning and Emperor himself saved the life of a Rajput woman. [10] While discussing about religious philosophy Abul Fazl criticized those religious leaders or preachers who only tried to establish received assertions and demonstrated no desire to arrive at truth by reasoning. [11] Abul Fazl also criticized bigots who tried to undermine the teachings of other religions not by logic but by bigotry. Emperor Akbar himself opined that, I realize that mere professing (of religion) is not enough, it can be beneficial which we select with clear understanding, mere practice can avail nothing in the sight of god. [12]

A glimpse on the idea of romance and love in Ancient and Medieval India also reflects the reigning ideas in the cultural life. In *Sakuntala*, poet Kalidasa very aptly depicted the state of mind of King Dushwanta, deeply in love with Sakuntala and missing his beloved. King Dushwanta imagined "love" as his only friend as if standing beside him and he lamented to his imagined friend;[13]

In vain, Love, did I make you big and strong,
Feeding you attentively with a thousand thoughts and feelings:
You drew your bow back to your ear
And I was the one you shot your arrow at.

This is definitely the expression of a very deeper, poetic, creative and sensitive mind. These lines of Kalidasa establishes the eagerness, romance of a lover who is always missing his beloved and thus becoming more eager to meet his beloved only to express her that he is waiting only to get a glimpse of her. In these lines there is no exaggeration, no dramatic expression, and no excess but this simple expression gives the idea of the nature of a consummate romance. Fayyazi, the poet during Akbar's reign wrote, "My heart is the world, and its Hindustan is initiated in the rites of idolatry and the rules of idol making" [i.e., my heart contains wonderful things]. [14] The poets were romantic indeed. Mazhari wrote, "I cannot understand the secret of Salma's beauty; for the more you behold it, the greater becomes your desire". [15] This is again an expression of a romantic mind. In fact this expression of Mazhari whereas romantic it is sensitive as well. The poet expressing the desire to know a woman, her self and the "greater desire" is not commodifying women as a body. Hussain Shah's general Paragal Khan's son Chutti Khan played a significant role in de-Sanskriting the famous epic *Mahabharata* to bring it more closer to the masses. It was he who told his court poet Srikara Nandi to narrate *Mahabharata* in Bengali instead of Sanskrit. [16] As Srikara Nandi summarized this; [17]

Samskrata Bharate Na Bujhe Sarbajana
Mora Nivedana Kichu Suna Kavigana.
Des'i-bhase ehi katha koriya pracara
Sancarau Kirti more jagata bhitara.

Translated version

Only few understands Sanskrit in India
Poets, let me to suggest humbly something.
Narrate the tale in local language
Tell everybody about this work of mine.

During the 17th century many poems were written by Muslim poets which imitated the narrative pattern of the Hindu poets. Sukumar Sen has observed that new type of poetry emerged in Bengal which were came to be known as *Satyapir Panchali,* [18] an amalgamation of the Hindu and Muslim tradition and this genre was actually an attempt to bring more closer the Hindus and the Muslims. Before the introduction of printed books in the colonial period, it appears that there were some 40 works composed in Bengali in the pre-colonial period. The most notable among them were *Chaitanya Charitamrita* by Krishnadas Kaviraj (1557), *Mansa Mangal* by Khemananda, *Dharma Gana* published by order of Lau Sen, *Subhankara* and *Guru Dakhina; Chandi* by Kabikankan, and *Annada Mangal* by Bharat Chandra.[19]

From this brief survey of pre-colonial manuscript book and literature we can locate few significant things. *Firstly,* apart from Sanskrit and Persian there existed a huge canon of regional literature like, Bengali. *Secondly,* the distinct *mode* of the literature was not exclusion and marginalization but inclusion. Nor does this mode promoted blind faith, despotism or narrowness. What is very much visible is that the intellectual leaders and also the rulers themselves exhibited concern for people, specially peasants and their well being. The rulers were spiritual in mind but they have not pursued a policy of promoting specific religion by thoroughly suppressing the other religions. At a later period we can see that even the received assertions of religions were rationally questioned, criticized due to the active encouragement of the rulers. The most notable thing is that there existed the idea of what should be an ideal state and what should be the characteristics of an ideal ruler. From ancient times down to the medieval times the ideal king has been put forward as one who should not be despotic, highhanded and tyrant. The ideal king implied a king who should look below, should give priority to the subject people, should think about their welfare and should not give priority to luxury, pomp and grandeur. These are not the signs of despotism or superstitious belief. What is most noticeable is that there existed no exclusion of ancient traditions in the medieval period. Rather we can see that there was an effort to assimilate the ancient "high" traditions within the indigenous society by relieving those from their "high" positions. For this reason Chutti Khan emphasized about spreading *Mahabharata* among the entire society by de-Sanskriting it and Abul Fazl emphasized that a king should not be a despot, irrational or narrow minded. *Thirdly*, another very significant feature of the literary *mode* was its attempt towards the creation of a secular literature. Although more or less the entire literary canon of pre-colonial India was deeply embedded in spiritualism, that spiritualism by its nature was not exclusionary, rather as in the most pronounced form in the *Pir* literatures the secular and spiritual was entwined with each other. The most notable thing of this *mode* was that be it Persian or Bengali there was no tendency of showing the different religious community as wide apart from each other with irreconcilable existence. *Finally*, the literary texts exhibits the signs of sensitive minds, be it romantic, like the one of Mazhari or romantic embedded in spiritualism as in the *Vaishnava* lyric songs and thus were in every sense contained the higher forms of art.

The Colonial Conquest and the introduction of printed book

The colonial rule started in India from 1757 and in the subsequent period it consolidated itself. Printed books about India dealing with India's history, society, economy and culture started with the introduction of the colonial rule itself. It is not possible to ascertain exactly when printed books started to arrive in India. In April of 1763 the Fort William Council requested the Court of Directors in London to send a collection of treaties and the Court reported the Council that they know about such book in French.[20] Subsequently in February 1765 in its letter to Fort William Council the Court reported that they dispatched *Histoire des Traites de Paix et Autres Negotiations*, in 18 volumes, folio printed at Amsterdam, through the ship Grenville to Calcutta.[21] These type of printed books were definitely meant for British East India Company servants. Subsequently, and more at a later period a large number of English printed books about India started to be published in London and then entered in India. For example, Luke Scrafton's *A History of Bengal Before and After Plassey* in 1763, J.Z. Holwell's *India Tracts* in 1764, *Memoirs Relative to The State of India* by Warren Hastings' in 1786 and many others. The output continued to increase in the 19th century which saw the *History of British India* by J.S. Mill in 1817, *History of Bengal* by Major Charles Stewart in 1813, F.J. Shore's *Notes On Indian Affairs* in 1838 and several other memoirs and tracts. With that started the publication of journals like *The Calcutta Review, The Friend of India, Bengal Hurkaroo* and from 1820 onwards several vernacular journals and English journals by Indians. It was through the newly rising educational set-up accompanied by newly founded schools and later colleges from the first quarter of the 19th century started the journey of the printed book in India. The Hindu college was established in 1817. D.P. Sinha has observed that by 1815 the Srirampur missionaries established more than 100 schools in different parts of the province of Bengal. [22] In 1817 *The Calcutta School Book Society* was established and D.P. Sinha has noted that by 1821 it published 1,26, 446 books in various subjects. [23] Promoting books in English became the cornerstone of the policy of colonial government. In his note of June 1824 A. Stirling observed that publication and preparation of books should receive a considerable attention from the government. [24] Not only that. A considerable section of the colonial officials were wanting not a mere spread of English language but a greater skill and proficiency. In fact in his letter of April 1824 Thomason proposed that annual examinations should be conducted in the Town Hall to ascertain the amount of proficiency in English. [25] Thus it appears that along with the introduction of English language and English books among the Indians there started an effort of *Englishing* the Indians. Subsequently, a decade later in his minute of February 1835 Macaulay observed that indigenous literary creations teach superstition, false history and so the indigenous literature has been pointed out as valueless. [26]

By the 1830's the English rage became so widespread that in his minute Macaulay noted that the School Book Society at that time was selling 7 to 8 thousand books per year in English, realizing a profit of 20% and on the other hand books printed in Arabic and Sanskrit remained piled up and unsold. [27] This trend continued to flourish in the coming decades. In fact sometimes total number of books published in English and Bengali stood very near to each other. The note on the State of Education in India in 1865-66 mentioned that the Director of Public Instruction in Bengal observed that learning English became so prestigious that "to secure for their children a knowledge of our tongue (English-K.C.) is the one object for which, as a rule, the people are willing to pay, and for this they will frequently incur an expense which would even altogether be disproportioned to their means".[28]

The School Book Society continued to flourish. Whereas in 1821 it published 1,26,446 books in the post mutiny period down to the 20th century the number of publication and the profit of the School Book Society rapidly increased as the chart shows;[29]

The Growing Performance of the School Book Society

Year	Number of Copies	Value (in Rs)
1865	1,84,043	64,317
1891	2,17,913	83,755
1906-07	-	1,52,348

The number of copies published by the School Book Society in different languages in 1865 was as follows;[30]

Publication of School Book Society in 1865

Language	Number of Copies
English	68,525
Bengali	83,588
Hindi	3,890
Oriya	12,824
Santhali	3
Khasia	511
Arabic	29
Persian	71
Urdu	2,683
Sanskrit	2,068
Anglo-Asiatic	9,851

The Bengali school books later became school books in different parts of eastern India. In 1878-79 the school Inspector of Behar circle reported that, "My predecessor formed the idea of getting Bengali school books translated into Hindi for the use of Behar schools. He brought out *Bidya-Ke-Neo*, a translation of the Bengali *Bodhodaya*". [31] Similarly the Joint-Inspector of Orissa reported that, "There is hardly any school book in Bengal worth the name on subjects other than literary which does not get itself translated into Ooriya in some shape or other". [32] Below is given the list of the books which were translated from Bengali to Hindi for Bihar's schools; [33]

Desi Lekha Jokha, compiled by Shyam Behari Lal

Vidya Ke Neo, Bodhodaya of Vidyasagar

History of Subah Bengal, from Bengali of Raj Krishna Mookerjea

Bostu Bichar, from Bengali *Bostu Bichar* of Ram Gati Nyaratna

History of India, from Bengali by Ram Gati Nyaratna

Sharir Palan, from Bengali by Jodunath Mookerjea

Ancient History, from Bengali by Bhoodeb Mookerjea

Natural Philosophy, from Bengali by Akhaya Kumar Dutta

Physical Geography, from Bengali by Radhika Prasanna Mookerjea

Khetnap Vidya, in Kaithi characters compiled by Gopi Lal

In Bengal a *Central Text Book Committee* was formed in 1875 and in 1890 it consisted of 17 members with Babu Bhudeb Mukherjee as President. The Text Book Committee was divided in 6 sub-committees of English language and literature, Bengali language and literature, Sanskrit, History and Geography, Mathematics and Science. The book publishers used to submit copies of books to the text book committee. The Secretary or the Assistant Secretary of the committee then used to forward to each member a copy of the book "with a covering letter, asking him to report on the book, in regard to its matter, manner, style, language, grammar, spelling, religious, political and moral tendency, paper, printing and price, and to state whether the book is recommended for school use or for prize or library purpose, and the school classes, if any, for ahich it is suitable". [34] In 1891-92 the text book committee reviewed 476 books for high and primary schools in which 152 books were selected for school, 82 for prizes and school libraries and 233 were declared unsuitable. [35] There was a local branch of Central Text Book Committee in Bihar also in the 1880's whose function was to review and approve text books. [36]

Apart from books for schools and academic world there emerged also a voluminous number of journals, newspapers and different other type of books in the late 19th century. It appears that during 1870 the India Office in London was negotiating with the *W.H. Allen & Co.* publishing company for supplying books to India Office and to India. From the letter of Grant Duff written in November 1870 to the *W.H. Allen & Co.* it becomes clear that *Allen & Co.* was ready to supply all books with few exceptions at a discount of 25%. At this rate the company was ready to pack the books, to get boxes for packing and also to bear the freight and shipping expense. In his letter Grant Duff informed the company that the Secretary of State for India in Council accepted the offer of the company. [37] The Secretary of State also directed that India Office will purchase all required books from the *W.H. Allen & Co.* publishing company. [38] It appears that at a later period further negotiations were going on with the company. In his letter of January 1872 the Assistant Secretary of Financial Department reported to the Secretary of the Government of

Bengal that the "Secretary of State would be asked to detail the more favourable rates at which M/S Allen & Co. were prepared to supply books for India". [39] By 1880's there were several Bengali and English newspapers. The leading Bengali newspapers were *Navavibakar, Sahachar, Anandabazar Patrika, Sulabh Samachar, Bangabashi, Sadharani, Somprakash, Dhaka Prakash, Bharat Mihir, Burdwan Sanjibani* and *Charuvarta*. The leading English newspapers were *The Englishman, Indian daily news, Statesman and Friend of India, Hindu Patriot (Hindoo Patriot), Indian Mirror, Reis and Rayyat, Indian Planters' Gazette, Sporting news, Darjeeling news, East and Bengal Times*, and *Behar Herald*. [40]

In September 1870 the Home Department passed a resolution. The resolution stated that books published in several Presidencies and Provinces should be classified in two types; "The first division should be according as the books are in English or in the various vernaculars of the country, and a sub-division should shew the following classes of books",

1. Original Works
2. Translations
3. Re-publications

Another subdivision should classify according to subject; Biography, Drama, Fiction, History, Language, Law, Medicine, Miscellaneous, Poetry, Politics, Philosophy, Religion, Science (Mathmetical & Mechanical), Science (Natural & Other), Voyages and Travels. [41] As a result in the subsequent period the Administration Reports used to publish annual report of publication of books in the Presidencies according to the aforementioned classification. In the coming decades several types of books were published and their number continued to increase as the chart below demonstrates; [42]

Category of Book	in 1882-83 Numbers	in 1894-95 Numbers	in 1904-05 Numbers
Biography	9	21	34
Drama	76	37	78
Fiction	35	68	-
History	26	47	99
Language	124	232	397
Law	33	43	-
Medicine	21	42	-
Miscellaneous	130	140	-
Philosophy	5	5	27
Poetry	108	112	207
Politics	-	3	-
Religion	96	194	815
Science Mathematics	39	48	109
Natural	16	20	-
Travels and Voyages	-	1	5

In 1906-07 total number of books published was 3,440 which became 4,093 in 1914-15 and in 1921-22 the number was 4,229. [43] Although books published in Bengali and other Indian languages sometimes dwindled in number but the number of books published in English witnessed generally a steady increase as the chart shows; [44]

Number of books published in different languages in different years

Language	in 1886-87 Number	in 1896-97 Number	in 1900-1901 Number
Bengali	979	752	910
English	220	263	286
Assamese	18	-	14
Oriya	91	156	158
Hindi	110	-	81

Colonial Printed hegemonic book and Civility Paradigm From Above

The colonial conquest of India through the winning by the British East India Company in the battle of Palashi in 1757 initiated a new era in the history of Bengal as well as India. In 1760 the Bengal Nawab Mir Kasim gave the East India Company the revenue collection duties of the districts of Burdwan, Midnapur and Chittagong. From then onwards, more after 1765 and systematically from 1772 onwards down to 1793 the British East India Company learnt, experimented and finally established an entirely new land revenue settlement in India. As a new land revenue settlement the Permanent Settlement of 1793 which created private property in land and subsequently established and empowered the new *Zamindar* class also necessitated to introduce new forms of coercion. For this reason, in order to protect the new land revenue settlements the East India Company also established from 1793 onwards a new network of police and prison. Starting from the first days of colonial rule down to 1793 and in the subsequent period the new land revenue settlement which corresponding to the empowerment of *Zamindars* continuously disempowered the peasants resulted in recurrent peasant resistance in every part of Bengal. The main reason was that the new rulers and their new institutions were an abrupt break from the pre-colonial period and hence even the coercive organs also finally remained incapable to ensure the required *legitimacy* to the new rulers. For the new rulers it became imperative to establish their *legitimacy* to rule for which it became necessary to *de-legitimise* the pre-colonial past.

It was on the background of this critical situation that the journey of printed books in India started. It was the printed books and tracts which started to give the shape of a new vision about the pre-colonial past and the colonial present.

This effort started from early 18th century when the English and French were combating with each other in order to establish their hegemonic control over India, already weakened due to the decline of the Mughal empire. To legitimise colonial rule it was necessary to legitimize colonial occupation and to de-legitimise the pre-colonial history. Thus, Luke Scrafton, an official of the East India Company in Bengal during the battle of Palashi observed in his book, A History of Bengal Before and After Plassey completed in 1763 that Bengal's Nawab Siraj-ud-Daula was a "capricious tyrant", "rapacious", "prodigal" and revengeful, cruel etc. [45] The obvious message here is that the East India Company liberated a country which was ruled by a *capricious tyrant* like Siraj-ud-Daula. The vast amount of literary output introduced significant new things. For example, in his *History of Bengal* (1813)Major Charles Stewart consulted a large number of Mughal and Sultanate period Persian sources and gave birth to a well researched book. But, there was a new paradigm in it. While discussing about the rule of the Bengal Nawab Murshid Kuli Khan, Stewart after narrating the early years of Murshid Kuli's rule observed that, "Having in the preceding pages, given what may be termed the dark side of Moorshud Cooly Khan's picture, we have now much pleasure in reversing the object; although we fear the portrait will still not be approved of by Englishmen". [46] While discussing about the political, economic and commercial condition of early 19th century British India in his book in 1832 Robert Montgomery Martin observed that, "India was subject to severe atrocity for countless age and this would have turned it into a howling wilderness........acquisition of the Indian empire by Great Britain has been the greatest blessing ever conferred on that land". [47] This was new and a thoroughly different *mode* from the pre-colonial period. As

Ismail Talib has observed that the introduction of English brought new literary *genres* and styles in places where it was introduced. [48] This became more demonstrative in J.S. Mill when he observed, "We have....seen, in reviewing the Hindu form of government, that....despotism and priestcraft taken together, the Hindus, in mind and body, were the most enslaved portion of the human race". [49]

This *mode* had its origin in enlightenment Europe. In the idea of "World History" or "Universal History" put forward by Kant, Herder, Condorcet or Buckle by "World History" enlightenment Europe meant the history of the nation-states of Europe. In that scheme of "World History" the nations of Asia and Africa has been perceived as outside the "World History" or in the stage of pre-history. In England this idea received a more matured form through the Shakespearean tradition which shaped the aggressive form of English national culture. The dramas of Shakespeare although composed in the 17th century but Shakespeare's drama, the *English* of Shakespeare gradually shaped up the national and cultural traditions of England. This gave birth to a specific *English* and at the same time suppressed the other cultural traits of the British isles. Willy Maley has observed that the Englishing of the British state happened due to Shakespeare which established *English* as the dominant language of the British isles.[50] The Englishness which originated from Shakespearean tradition not only influenced national life in England but it played a crucial role in the formation and consolidation of British colonial empire, chiefly India. It was Shakespearean tradition whose application in British colonial empire not only influenced historiography but shaped up the hegemonic culture. Ania Loomba has observed that in order to legitimize occupation English nationalism relied on cultural distinction. [51]

The Tempest is a drama of William Shakespeare. The principal characters of this drama are *Prospero, Caliban, Miranda, Ferdinand, Ariel, Antonio, Gonzalo* and others. Prospero was duke of Milan who lost his Dukedom due to the conspiracy of his brother Antonio. Prospero became compelled to live in a remote island with his daughter Miranda and the other living being of the island was Caliban who was a subhuman monster. Prospero possessed magical power through his books and became able to control spirits like Ariel. Prospero turned Caliban as his subject and became the ruler of the island. Later Prospero's art compelled his brother to repent and he got back his kingdom. His daughter became the betrothal of Ferdinand. One principal component of the drama was Caliban's subjugated position and the attitude of Prospero and all others towards Caliban. *The Tempest* became a cultural weapon which legitimized conquest, occupation and subjugation of foreign land by England and thus served expansionist commercial and economic purpose of England. For this reason Luke Scrafton and many others tried to justify the British occupation of Bengal in 1757 by terming the Nawabi rule as despotic and oppressive rule. In *The Tempest* when Caliban is admitting that Prospero with his art can even control Caliban's god Setebos [52] it indicates that in the eyes of the English "We" even the gods of the "Other", in this instance Indians are powerless and non-entity and can be subjugated by the English "We". This becomes evident when Caliban was telling Prospero, [53]

> *And show'd thee all the qualities O' th' isle,*
> *..............................and here you sty me*
> *In this hard rock,................................*

Martin Orkin has observed that *The Tempest* suggests to the colonizer or the traveler the realization of an ordering presence in the new land. [54] Further, Terence Hawkes has observed that what was meaning to the Asian/Indian "other" was meaningless to the European/English "We" and a subjugation of the Asian meaning by European meaning implied sense, order, money. [55] Thus what was history and culture in a pre-print society became pre-history and savagery in the print culture introduced by the colonial rule with a new *mode* originated from the Shakespearean tradition. *The Tempest* emphatically establishes that Europeans/English are in every respect learned and superior. For this reason Caliban while thinking about attacking Prospero gave priority "First to possess his books and to burn but his books". [56] Thus under colonial context the entire canon of pre-print manuscript book became "charm" like the knowledge of Sycorax and printed book of Prospero became symbol of "Wisdom". [57] It was with this new *mode* that the book in colonial India was introduced.

It appears that Robert Darnton was discernibly correct when he observed that books do not merely recount history, they make it. [58] In colonial India when the printed books started to entrench in the indigenous society with its own genre of Shakespearean tradition whose benchmark was "exclusion" of everything pre-colonial by terming them as "despotic", "irrational" and "dark" it soon gave birth to a new culture. In fact, the book played a very significant role in shaping up of a hegemonic culture. In 1823 The Secretary to the Calcutta Madrasa, Lumsden proposed for changes in the primary text books. Lumsden proposed that books should be modeled in such a way which will contain those things which are the signs of European taste and will reject those which will appear as objectionable or erroneous according to the European taste and he further proposed to cultivate a taste for English literature. [59] From this time onwards the newly introduced *mode* with its background of

Shakespearean tradition has started to become disseminated more methodically and thoroughly.

In the post 1820's, as Sukanto Chaudhuri and Ananda Lal has observed that the Bengal Zamindars and the rising intelligentsia started to devote a lot of time in learning and becoming "English" thorough Shakespearean recitation to full scale productions.[60] From the very beginning the colonial government's policy was to create a core group of Indians who will imbibe the Shakespearean Weltanschauung and they will then play the role of hegemonising the cultural world view which will help the Company Raj to thrive. The Permanent Settlement created zamindars appeared to the Company government as the most suitable section of the society to perform the task. Governor-General Auckland's private Secretary J.R. Colvin observed in his note in 1839 that improvement in education was required for the higher classes and this in its turn can bring "beneficial change in ideas and feelings of the community".[61]

It was through the Shakespearean drama that the English national culture penetrated in the Indian society. The literary canons of Shakespeare were introduced in the school curricula of the early 19th century Bengal. The newly rising group of the educated youths of Hindu College, Oriental Seminary etc were encouraged to study, recite and act the plays of Shakespeare. In order to develop a taste and appreciation for the Shakespearean drama the colonial government supported the ideas of discouragement and blackening of the India's own dramatic tradition. As *The Calcutta Review* of 1851 observed that ancient India have a genius dramatic tradition but it reflects polytheism and superstition.[62] In this category came notable ancient dramas like *Mrichchakatika, Sakuntala, Malati and Madhava, UttaraRamcharita, Vicramorvasi,* etc. The Calcutta Review not only argued Indian dramatic tradition as a degenerated one but put forward the English drama as a "Superior" substitute, because Indian dramas were declared as "Worthless with indecent songs and gestures". [63] Although Shakespeare's dramas were played in Calcutta from 1780 onwards but initially it remained confined within the British community. From 1820's started the participation of the newly educated Indian youths in the form of recitation, short scenes etc of the poems and dramas of Shakespeare. *The Calcutta Gazette* of 17th January 1828 highly praised the youth of Calcutta for embracing Shakespeare [64] and thus followed the process of imbibing the Shakespearean ethos and reshaping of the Indian culture according to the needs of the Raj. Very rightly, Terence Hawkes has observed that Shakespeare is a powerful ideological weapon, always available in the periods of crisis, and used according to the exigencies of the time to resolve the crucial areas of indeterminacy. [65] The Bengal Zamindars started to take an active role for the promotion of the Shakespearean drama. *The India Gazette* of 14 September 1831 highly appreciated a meeting organized by Babu Prasonno Kumar Tagore to establish a theatre. [66] In the coming decades more and more youths started to practice recitation, acting etc of Shakespeare's plays. On 5th September 1853 *Othello* was performed in Calcutta by a troupe of Indians who were the former students of Oriental Seminary. [67] The Indians acted under the Supervision of former Sans Souci Theatre personnel Mr. Clinger. [68]

Othello is a drama of William Shakespeare. The principal characters in this drama were *Othello, Desdemona, Cassio, Iago, Brabantio, Emilia* and others. *Othello* was a black skin very able military general of Venice. *Othello* used to narrate his several stories to Desdemona, the daughter of senator Brabantio. Gradually *Othello* became a hero in the eyes of Desdemona who fell in love with *Othello* and both married.

Old military officer Iago disliked *Othello* and *Othello*'s close and trusted lieutenant Cassio. Due to Iago's machinations there arose jealousy in simple minded *Othello* because he imagined about an affair between his wife Desdemona and Cassio. This finally resulted in killing of Desdemona by *Othello* and then suicide by *Othello*. The drama appears as a story of jealousy between men regarding women but the drama portrayed *Othello* as a jealous, dull and revengefull black skin lover. *Othello*, the black Moor (Muslim) has been for a long time a matter of controversy. Even, Maryland ones described *Othello* as a white man. [69] But *Othello* has been presented as a black by Shakespeare and this have a profound significance. As Ania Loomba has observed that a serious consideration of the race relations within the play blurs the theme of sexual jealousy between men regarding women as an universal phenomenon. [70] On repeated occasion the other actors termed *Othello* as a black with special stress and on other instances the drama establishes the black stereotypes of dullness, foolishness, lack of understanding capacity and reasoning, monstrosity and revengefulness. Within the drama, as Ridley has suggested that *Othello* is completely master of himself when he coldly decided to take revenge on Desdemona. [71] Nowhere is the black stereotype more explicit than when *Othello* was lamenting after discovering his fault; [72]

> *Of one that lov'd not wisely, but too well:*
> *Of one not easily jealous, but being wrought,*
> *Perplex'd in the extreme; of one whose hand,*
> *Like the base Indian, threw a pearl away,*
> *Richer than all his tribe:*

In the most tragic moment of the drama the tragic hero is declaring that a pearl / white woman is not and can not be an equal to a base Indian whose superiority in every respect is above the tribe of the Indians. The base Indian can love, but not wisely and thus a base Indian's love becomes equal to the love of a beast who lacks the fine, poetic and romantic sense of love. When we keep in mind the entire play and on different occasion the application of the black stereotypes then the whole thing becomes more clear. When Emilia is cursing Othello by saying,[73]

O' the more angel she,
And you the blacker devil!

It, in the dramatic moment largely emphasizes the stereotypes of a "black devil"/non-European other/ Calibans and *Othello's* guilt becomes entwined with the blacks. Again when Emilia is crying "O thou dull Moor", or when Iago was saying that a Moor like *Othello* can "tenderly be led by the nose .... as Asses are", "an old black Ram", "the devil will make a grandsire of you" all affirms and establishes the picture of a black stereotype who is full, insensitive, dull and devil.[74] It becomes clear that a drama like *Othello* acted by the black Indians in front of a black audience will give booster to the idea of a *white/European/English* cultural superiority in comparison to a *black/Indian* cultural inferiority. Naturally, culturally inferiors will try to identify themselves with the culturally superiors. Gauri Vishwanathan has observed that a vital connection (may be subtly) exists between a discourse in which those who are to be educated are represented as morally and intellectually deficient and the attribution of moral and intellectual values to the literary works they are assigned to read.[75] Thus, in the Indian colonial context Shakespearean tradition being introduced through printed books with its own distinct mode more firmly established a demarcation line between *culture and unculture, civil and uncivil, god and devil, educated Zamindar cum intelligentsia/ Prospero and rebel peasants/ Calibans.*

The new mode pervaded the entire print output, be it English or vernacular. The *Hindoo Patriot* gladly applauded in May 1854 the theatre rage in Calcutta as a sign for "better" because talented Bengali youths were giving a new life to Shakespeare. [76] The new genre promoted and encouraged the exclusionary vision which directly or remotely always contrasted the pre-colonial past and anti-colonial thoughts as "irrational" or "despotic". In 1832 *The Samachar Darpan* termed Titu Meer's rebellion as Muslim fanaticism. [77] *The Samachar Darpan* also argued that those who speak contemptuously about the colonial rule and the attendant changes and still have attachment to old customs were foolish persons. [78] In 1855, to the *Hindoo Patriot* the *Santal* rebels appeared as people who inspite of long colonial rule still remained savage. [79] In its next issue the *Hindoo Patriot* criticized the government military for not becoming able to suppress the rebels. [80] The *Hindoo Patriot* later proposed for opening a relief fund for the *Mahajans* and planters who were driven away by the *Santals*. [81] During the great Indian revolt of 1857 The *Hindoo Patriot* commented that the Asiatic idea of rebellion means killing and capturing the capital of the king. [82] In its issue of 28th May 1857 the Hindoo Patriot highly applauded the demonstrations of loyalty by the intelligentsia to the British rule. [83] In fact, during the revolt of 1857 one serious concern of the Bengali intelligentsia was how to assist the state to confront a "brutal.....body of ruffians". [84]

Hiren Gohain has observed that the Bengal intelligentsia under British tutelage used to consider them as "sobhyo" or civilized according to the English paradigm of civility. This "sobhyota" or civility of the intelligentsia kept itself in distance from the working masses whom they considered as objects of charity but not fit to become masters of their own destiny. [85] At the end of the year 1857 The *Hindoo Patriot* commented in its December issue that the rebellion of 1857 was an anarchy which was a regular feature of the Mughal rule.[86] When torture by the police in custody became a serious issue in Bengal, the *Hindoo Patriot* proposed in June 1855 that a remedy to the problem was a more closer and direct supervision by Europeans. [87] At a later period when Bankimchandra Chatterjee wrote *Anandamath,* there, the character of *Sannyassi* Jibananda commented that the English should rule the country and that rule will bring progress for the country. [88] To Rabindranath also the pre-colonial, specially the Sultanate and Mughal period appeared as a period of anarchy and not forms a part of the history of India because it was not the history of motherland and in his "India" or "Bharatbarsha" the indigenous Adibasis were the outsiders/others. [89] Apparently although it appears that Tagore tried to question the remodeling of India's past under colonial rule but while doing so he placed himself within and not outside of the Shakespearean tradition.

In her *Sentinels of Culture* Tithi Bhattacharya has observed that the pre-colonial education system was exclusionary because it was based on caste system and the western education was a capitalist enterprise and it was a democratic education. Bhattacharys further observed that between pre-colonial orality and with the introduction of print under colonial rule there happened no rejection of indigenous tradition but a rearticulation.[90] While discussing about the coming of the book in India Anindita Ghosh has observed that since before the coming of the printed book in India there already existed a pre-colonial literary tradition which established clear taste and preference of reading so when the printed book arrived in India under colonial rule the reading practice has not changed.[91] Ghosh concluded that since there existed multiple reading practice so the printed text could not play any significant role in acculturing its Indian readers and thus the book in colonial India failed to disseminate a civilizing message from the part of the colonial rulers.[92] Similarly, while discussing about the history of the book in India Priya Joshi observed that India under colonial rule gives the image of a situation where both the colonial rulers and the Indians stood with outstretched arms facing each other, each side taking, exchanging from each other and thus there was no exclusionary practice.[93] Further, Priya Joshi also observed that the political effects of a text lie outside it in its consumption by readers and so it is justified not to locate political meaning solely in the text and Joshi in fact tried to establish the readers as producers of text.[94]

After going through the entire aforementioned book history of India from pre-colonial period down to the colonial period it becomes clear that the observations of Anindita Ghosh and Priya Joshi are misleading and gives a wrong idea about the book history of India. What we have already noticed is that the history of book from pre-colonial to colonial period or rather from the pre-print manuscript book to the printed book was not a history of continuity, but change. There happened nothing like exchange, sharing or inclusion of ideas between Indians and the colonial rulers under the colonial rule. What happened was a thorough and abrupt change of *genre* from pre-colonial to colonial period. The new *genre* in colonial India originated from the Shakespearean tradition and by its very nature this genre promoted marginalization and exclusion of the pre-colonial indigenous genre. At a later period with the passage of time this new *genre* shaped the literary output of India which is visible in the published works of Bankimchandra, Tagore, in the journals like *Hindoo Patriot*, *Samachar Darpan* and in fact the entire print output of 19th and 20th century India is replete with this *genre*. Not only that. The most interesting thing is that books in Bengali were basically an adaptation of Shakespearean tradition from the English books. In his report of 1878-79 the Inspector of Behar Circle reported that Hindi books which were translations of Bengali books were "themselves compilations and adaptations from English works".[95] Thus there came into being a long chain of dissemination of Shakespearean tradition. It appears justified to argue that exactly this was the motive of the colonial government from the very beginning. In his note of June 1824 A. Stirling has observed that the purpose of government fellowships in India should not be creating scholars. [96] Stirling further observed that fellowships should result in a good grasp of English language so that "selected English works" can be subsequently translated by Indians in Indian languages.[97] Further, what Priya Joshi has suggested is going outside the text and thus abandoning the text to locate textual interpretation. While it is true that readers read and interpret the text but that does not implies that readers becomes the producer of text. Readers interpret the text, as Mccleery and Finkelstien has observed , on the basis of social experience, shared value and cultural references.[98] So, it appears logical to argue that to understand and discuss reading the text should be given first priority. To understand racism and affirmation of black stereotype in colonial India it is necessary to understand and interpret *Othello* and not to seek for a reader of *Othello* as the producer of *Othello*. Terry Eagleton has observed that no 'text' can be independent, it subsists in its complex relation to other uses of language, and so to ideologies and social formations.[99] Eagleton concluded that a critique must question the eloquent silence, should explain the ideological necessity of its 'not-said' because it is in the eloquent silence, in significant gaps and fissures of a text where the ideology of the text is present.[100] As we have already observed that Major Charles Stewart in his *History of Bengal* asked the readers whether Nawab Murshid Quli can at all be approved of by Englishman as a good ruler. It is in this unsaid words, in the silence after the question which will *compel* a reader studying, reciting and acting *The Tempest* or *Othello* under *colonial educational curricula* to conclude that the very "Englishness" of Prospero was absent in Murshid Quli and what was present was a despot "black devil" like *Othello* who was "capable of all ill" and "deserves more than a prison" like Caliban as Miranda expressed in *The Tempest*.[101]

It was only the resistance to colonial rule which gave birth to an opposite style and genre in indigenous literature. As is evident in the poem of Sajan Gaji who himself participated in the Wahabi movement of Bengal under the leadership of Titu Meer. In his song the Gaji wrote that under the leadership of Ghulam Masum the rebels attacked the colonial rulers, the frieghtened rulers fled and thus ended the Company's Raj.[102] During the Santal rebellion of 1855 the rebel Santals sang songs whose message was that Hool or rebellion for the sake of motherland became necessary because the "trader thieves" robbed the land.[103] Similarly, in his Padmini Upakhyan Rangalal Bandopadhyay wrote that,[104]

"Who wants to live without freedom
Hey! Who wants to live?
It is heavenly pleasure a day of freedom"

There was no socio-cultural acceptance of the colonial rule by the peasant masses in rural India. The *Kol, Santals* or Sepoys and peasants never thought that the colonial rule was a rule progressive and civilized in character. The resolute expressions which upheld the Raj of the rebels, the end of the British Raj only establishes that rural peasants had no fascination or attachment to the "civility" which the *Hindoo Patriot* upheld. The popular culture never received appreciation from the intelligentsia because this culture always questioned the legitimacy of the colonial rule and thus was a purely opposite version of the constitutionalist culture/legalist culture of the intelligentsia. But, this does not indicates something like the *Subaltern School's* assertion that parallel to the dominant and hegemonic Shakespearean tradition there existed a separate and autonomous indigenous rebel *genre* of literature. Even, within the rebel *genre* the Shakespearean tradition sipped through as it becomes evident from the *Jager Gan* of Ratiram Das where after highlighting the courage of the Rangpur peasant rebels the poet assured that the country will prosper because the British are the emblems of justice. [105] Thus, the idea of a Prospero with his "Wisdom" even pervaded the very rebel thoughts who were supposed to question the hegemony of Prospero.

A new *genre* started to develop at a much later stage from the 1940's which became best manifested in the works of Manik Bandopadhyay. Manik Bandopadhyay observed in his *Why do I write,* that a writer is a labour who works with pen. [106] Further, he always adhered to higher forms of art, expression and language, but at the same time observed that even if literary production assumes propagandist form, in times of social change that can be regarded as a sign of literary progress. [107] What was excluded under the colonial rule was indigenous history, knowledge. What was implanted from above through printed books was a culture and tradition conducive for the Raj. In 1846 while discussing about English literature in India *The Calcutta Review* observed that India was a less refined and savage society before the introduction of colonial rule because there was no book. [108] Caste system or religious distinction of pre-colonial period never excluded indigenous history, culture and knowledge. Secondly, in pre-colonial class divided society the indigenous *adibasis* were not treated as "others" and the state never interfered in their customary privileges. But, colonial historiography generated the view of *adibasis* as "others", the state interfered on their customary privileges and marginalized and displaced them.

The Book and Civility paradigm in Post 1947 India

On the eve of 1947 the biggest agenda in front of the post 1947 India government was to implement the development programmes. Development was required in all its aspects, economic, political and cultural. But, from its very inception the post 1947 government, as Smitu Kothari and Imtiaz Ahmed stated, continued to apply the same mode and methods of resource use which was applied by the colonial government for colonial control and resource use. [109] In the post 1947 India, specially after the 1960's with the growth of schools and universities in number, book production also increased enormously. With that enormously increased the output of journals, magazines and newspapers. From late 1970's onwards the book itself changed, because the quality of cover, paper and as a whole the book witnessed a sea change. At the same time increased the output of both the vernacular literatures and the Indian English literatures.

A continuation of the policies of the colonial state in the post 1947 India by its very nature was self-defeating to the development question. The continuation of the colonial mode of resource use naturally necessitated and rather kept intact the colonial perception of state and society. This could happen because absence of land reform resulted in continuation of semi-feudal landlordism where be it Maharashtra or Orissa or West Bengal everywhere the *Adibasi* people continued to be marginalized. This resulted in the continuation of the colonial culture also.

In the post 1947 India the secondary level school text books on history still continues to echo the same idea of civility which has been initiated under the colonial rule. The Brahmo Samaj, the Young Bengal etc movements are regarded as a "Renaissance". The argument here is that the colonial rule initiated a culture of questioning the received ideas of tradition, custom and culture. This is referred as rationalism. [110] Well known school text books argued that, "In the beginning of the 18th century in India there existed a situation conducive to backwardness which originated due to the downfall of the Mughal empire. It gave birth to superstition and dogmatism". [111] Another book argued, "The British East India Company assumed the authority of India at a time when India already turned to a degraded country in every respect. The cultural life was decadent.....But gradually in the enlightened circle of England there emerged the idea of introducing the Indians with the developed culture-civilisation and science of west". [112] Not only school text books. The idea of a pre-colonial dark age followed by a colonial civilized age continued to be manifested in the major and significant historical projects of post 1947 India. One of the most ambitious post 1947 project was the multi volume Bharatiya Vidya Bhavan series books edited by nobody less than R. C. Majumdar. R. C. Majumdar only reaffirmed the assertion of the dawn of a civilized state and society under colonial rule. In his work Majumdar observed that "The onslaught of Islam accompanied by a marked decadence of culture and the disappearance of the creative spirit in art and literature, seems to mark AD 1000 as the beginning of the medieval age". [113] As according to R.C. Majumdar the pre-colonial India was a Caliban's world so how India get rid of it? Majumdar answered, "The British rid India of the tyranny of the Muslims, Bengal renaissance was a true renaissance, wider deeper and more revolutionary than that of Europe". [114] R.C. Majumdar highly hailed the colonial rule as a period when India passed from *medieval* to *modern* age because British rule resulted in the intellectual development of the people. [115] There are more thoroughly misleading interpretations of Majumdar. As Vinay Lal has observed that, Majumdar agreed with the view that Indian nationalism was a consequence of introduction of western thought in India. [116]

This same idea is also present in the literary world which regularly receives its feedback from historiography. In this context Priya Joshi has observed that with the arrival of the *Midnights Children* of Salman Rushdie from the subsequent 1980's the landscape of Indian novel in English thoroughly changed and ushered in a new era, specially because it gave birth to an Indian English with indigenous link. [117] Salman Rushdie in his editorial note of The Vintage Book of Indian English observed that today Indian English literature is witnessing a considerably significant amount of output perhaps more significant than the total vernacular literatures of India. [118] Further, Rushdie highly applauded that Indian English by establishing itself as a distinct English created a space of its own. [119]

In one of the most significant chapter of *Midnights Children,* i.e., in *Midnight* while describing about the declaration of emergency of 1975 which itself is the most significant part of the chapter, Rushdie narrated, "When the constitution was altered to give the Prime Minister well-nigh absolute powers, I smelled the ghosts of ancient empires in the air…in that city which was littered with the phantoms of slave kings and Mughals, of Aurangzeb the merciless and the last, pink conquerors, I inhaled once again the sharp aroma of despotism". [120] After our entire study it leaves no doubt to conclude that this invoking of despotism is that same oriental despotism and its concept which was introduced under colonial rule through the application of the Shakespearean tradition. The emergency itself was a manifestation of the conflict between state and civil society in India and the implementation of emergency further intensified the conflict and widened the gap between the state and civil society in India. This state despotism, inherited from the colonial state could have easily compared with the role of the British Indian state in the year 1857. But to Rushdie, and very similarly like the concepts of Tagore, Bankim Chatterjee and many others despotism means pre-colonial history of India. Thus the pre-colonial has been put forward as a Caliban's rule so naturally the colonial and the post 1947 is bound to be a Prospero's rule in this vision. Moreover Salim Sinai termed the post 1947 India as *Kaliyuga* by invoking the Indian mythological tradition. [121] This is bound to generate a sense of lack, fate or tragedy. But since the pre-colonial and thus essentially the anti-colonial has been put forward as a Caliban's world, full of "charm" and no "wisdom" so naturally what will prevail and bound to prevail is the colonial and thus essentially the post 1947 nation state despite its despotism and the hapless surrender of the civil society to that state. Finally, it is also equally problematic to accept that a new Indian English has created a space of its own, with its own autonomy. The "Hinglish" which according to Rushdie is Indian English is essentially a "hybrid" creation and this hybridity has been put forward by Rushdie as something like a liberator or an emancipator. Ania Loomba has observed that "hybridity" does not automatically indicate resistance or agency. [122] Jonathan Dollimore has observed that in the contemporary times the literary model concentrates on the instabilities within the dominant, identifying for instance, ways in which the marginal is subversive of the dominant. [123] In this way subversion appears as an agency, an autonomous space, but, as Dollimore further observed that what is absent in this literary model is the intellectual commitment to engage with the cultural real.[124] The idea of hybridity is trying to establish a relative freedom, but in actuality the binary of oppressed and oppressor remains and thus the power-subjugation relation in a global scale remains intact and a further continuation of hybridity is bound to continue the Shakespearean tradition. Further, Viney Kirpal has observed that the assertiveness of hybridity lends an appealing quality to the works which promotes hybridity.[125]

So, from the entire abovementioned survey of the history of book in India we can conclude that, firstly, in the pre-colonial India there existed a considerable amount of manuscript book with its distinct mode. The main feature of this mode was that it was not exclusionary and it was born from within the indigenous society and its literary as well as historical tradition. Be it the court languages or high languages like the Sanskrit or Persian or the regional language like Bengali never perceived the indigenous as a distinct other self nor it tried to permanently detach it from the indigenous. Secondly, in colonial India the colonial rule for its own legitimacy and stability introduced the Shakespearean tradition accompanied by its own distinct mode. This mode promoted binary contrast and exclusion and detached the colonial present from the pre-colonial past which in its turn marginalized the indigenous history and literary tradition. On the other hand this mode promoted attachment towards the Raj as an embodiment of Prospero with his "wisdom" and power which as a whole is "English/civil" and a detachment or distance from that very "English" identity implied a sub-human identity like the fish monster Caliban with his all ugliness. Stuart Blackburn and Vasudha Dalmia observed that enlightenment strengthened the link with traditional literary creations in

Europe.[126] For this reason in Europe as a whole there never happened a discontinuity with indigenous tradition and history. But in India under colonial rule book was one significant tool which resulted in discontinuity with indigenous tradition and history. Finally, this mode is still existing in the post 1947 book where along with its continuing existence it now is drawing sustenance from post-modern ideas of memory, fragment, narrative, hybridity, etc. It leaves no doubt to conclude that the whole literary mode which originated in colonial India had to be abandoned otherwise literary production and hence book will serve only hegemonic interest.

List of Abbreviations used in the Endnotes

WBSA *West Bengal State Archives*, Bhawani Dutta Lane, Kolkata

RPI *Report of Public Instruction in Bengal* (relevant years mentioned)

Admn RP *Bengal Administration Report* (relevant years mentioned)

SEP benoy ghosh, *Selections From English Periodicals in 19th Century Bengal, Kolkata, 1978-1980* (relevant volumes mentioned)

SER *Selection From Educational Records, Part I, 1781-1839,* Kolkata, 1920, New Delhi,1965

SER EDU *Selection From Educational Records,* Vol-I, Educational Reports, 1859-1871, New Delhi, 1960

Notes :

1. Letter from Rammohan Roy to Victor Jacquemont, in A.C. Gupta Ed. *Studies in the Bengal Renaissance,* Kolkata, 1958, pp. 13-14, Gautam Bhadra, *Narkelberer Jang,* Kolkata, 1990, p. 59.
2. Kautilya, *Arthasastra,* R. Shamasastry (ed.), Mysore, 1967, p. 10.

3. Zia-ud-din Barani, *Tarikh I Firuz Shahi*, in The History of India, Vol-III, Ed, John Dawson, Allahabad, pp. 230-231.
4. Firuz Shah Tughlaq, *Futuhat I Firuz Shahi*, in Ibid, pp. 374-375, 388.
5. *Ibid*, p. 378.
6. Shams I Siraj Afif, *Tarikh I Firuz Shahi*, in ibid, p. 300.
7. Abbas Khan Sarwani, *Tarikh I Sher Shahi*, in John Dawson Ed. The History of India, Kolkata, 1871 & 1957, p. 146.
8. Abul Fazl, *Ain I Akbari*, Tr. H. Blochman, Ed. S.L. Goomer, First Published, 1871, Second Published, New Delhi, 1965, pp. 2-3.
9. *Ibid*.
10. Abul Fazl, *Akbar Nama*, John Dawson (ed.), Kolkata 1875 & 1953, pp. 72-73.
11. *Ibid*, p. 63.
12. *Ibid*, p. 64.
13. Kalidasa, Sakuntala, Act III 6, in Michael Coulson, *Three Sanskrit Plays*, Middlesex, 1981, p. 72.
14. *Ain-I-Akbari*, op. cit., p. 623.
15. *Ibid*, p. 654.
16. Sukumar Sen, *History of Bengali Literature*, Kolkata, 1992, p. 76.
17. *Ibid*, p. 355.
18. *Ibid*, pp. 145-146.
19. Early Bengali Literature and Newspapers, in *The Calcutta Review*, Vol-XIII, No. XXV, January-June 1850.
20. Letter from Court of Directors, London, to Fort William Council, Calcutta, 22 February 1764, in *Fort William India House Correspondence*, Vol-IV.
21. Letter from Court of Directors, London, to Fort William Council, Calcutta, 15 February 1765, in *Fort William India House Correspondence*, Vol-IV.
22. D.P.Sinha, *The Educational Policy of the East India Company in Bengal to 1854*, Kolkata, 1964, p. 40.
23. *Ibid*, p. 44.
24. *Note by A. Stirling, 20 June 1824 No.33, in Proceedings of the General Committee of Public Instructions, 14 January 1826, Correspondence and Proceedings, Vol-III, Part-I, Minutes of the Committee, 1823-1838 (WBSA).*
25. *Letter from Thomason, 6 April 1824, in Proceedings of the General Committee of Public Instructions, 14 January 1826, Correspondence and Proceedings, Vol-III, Part-I, Minutes of the Committee, 1823-1838 (WBSA).*
26. Macaulay's Minute, 2 February 1835, in SER, p. 115.
27. *Ibid*, p. 114.

28. Note on the State of Education in India 1865-66, in SER EDU, p. 272.
29. Note on the State of Education in India 1865-66, in SER EDU, p. 278, R P I 1891-92, p. 126, *Report of Progress of Education 1902-03*, p. 176.
30. Note on the State of Education in India 1865-66, in SER EDU, p. 278.
31. RPI 1878-79, p. 101.
32. RPI 1878-79, p. 102.
33. RPI 1878-79, p. 101.
34. RPI 1891-92, p. 123.
35. RPI 1891-92, p. 124.
36. Admn RP 1882-83, p. 489.
37. *Letter from M.E. Grant Duff to W.H. Allen & Co., India Office London, 4 November 1870 No. 394 , in Proceedings of the Lieutenant Governor of Bengal, Education Department, March 1872 No. 87 (WBSA).*
38. *Ibid (WBSA).*
39. *Letter from Hollingberry, Assistant Secretary, Government of India, Financial Department to the Secretary of Government of Bengal, 26 January 1872 , in Proceedings of the Lieutenant Governor of Bengal, Education Department, March 1872 No. 85 (WBSA).*
40. Admn RP 1882-83, p. 490, Admn RP 1885-86, p. 321.
41. *Resolution of Government of India in the Home Department (Public), Simla 20 September 1870, in Proceedings of the General Department, October 1870 No. 53 (WBSA).*
42. Admn RP 1882-83, p. 491, Admn RP 1894-95, p. 342, Admn RP 1904-05, p. 138.
43. Admn RP 1906-07, p. 128, Admn RP 1914-15, p. 126, Admn RP 1921-22, p. 292.
44. Admn RP 1886-87, p. 309, Admn RP 1896-97, pp. 334-335, Admn RP 1900-1901, pp. 218-219.
45. Luke Scrafton, *A History of Bengal Before and After Plassey 1739-1758*, London, 1763, Reprint, Kolkata, 1975, pp. 56, 102-103.
46. Charles Stewart, *History of Bengal*, 1813, Reprint, New Delhi, 1973, p. 405.
47. Robert Montgomery Martin, *The Political, Commercial and Financial condition of the Anglo-Eastern empire in 1832*, London, 1832, p. 19.
48. Ismail Talib, *The Language of Postcolonial Literature*, London, 2002, p. 82.
49. J.S. Mill, History of British India, in Patrick Brantlinger, *The Rule of Darkness*, Cornell, p. 78.

50. Willy Maley, 'This Sceptred isle': Shakespeare and the British Problem, in John J. Joughin Ed, *Shakespeare and National Culture,* Manchester, 1997, pp. 86, 96.
51. Ania Loomba, *Colonialism/Postcolonialism*, London, 1999, p. 71
52. William Shakespeare, *The Tempest*, Frank Kermode (ed.) Arden Edition, London, 1994, Act I Sc II 375, p. 33.
53. Ibid Act I Sc II 339-346, p. 31
54. Martin Orkin, Whose things of darkness, in John J. Joughin (ed.), *op. cit.*, p. 145.
55. Terence Hawkes, Swisser-Swatters, in John Drakakis (ed.), *Alternative Shakespeares*, London, 1986, p. 28.
56. *The Tempest*, Opcit, Act III Sc II, 90 & 93, p. 82.
57. Ibid, Act I Sc II 339-346, p. 31 & Act III Sc II 85-92, pp. 81-82 .
58. Alistair Mccleery and David Finkelstein, *An Introduction to Book History*, London, 2005, pp. 12-13.
59. Ibid, p. 50.
60. Sukanto Chaudhuri and Ananda Lal (ed.), *Shakespeare on the Calcutta Stage*, Kolkata, 2001, p. 7.
61. Extract from note by J.R, Colvin, private Secretary to Governor General, in Auckland's Minute of 24 November 1839, in SER
62. Hindu Drama, *The Calcutta Review*,1851, Vol-XV, No.XXIX.
63. *Ibid.*
64. Chaudhuri and Lal, *op. cit.*, p. 24.
65. Terence Hawkes, *op. cit.*, p. 43.
66. India Gazette,14 September 1831, SEP-III, p. 45.
67. Bengal Hurkara, 28 September 1853, in Chaudhuri and Lal, *op. cit.*, p. 26.
68. *Ibid.*
69. William Shakespeare, *Othello* ,M.R. Ridley (ed.), Arden Edition, London, p. ii.
70. Ania Loomba, *Colonialism/ Post Colonialism*, London, 1999, p. 84.
71. *Othello, op. cit.*, p. LX.
72. *Ibid*, Act V Sc II, pp. 345-349.
73. Ibid, Act V Sc II, p. 131.
74. Ibid, Act V Sc II, p. 226 & Act I Sc III, p. 399, Act I Sc I, 88, Act I Sc I, p. 91.
75. Gauri Vishwanathan, *Masks of Conquest*, New Delhi, 1989, p. 4.
76. *Hindoo Patriot*, 11 May 1854 SEP- III, p. 143.
77. Samachar Darpan, 26 December 1832, Titumeer's Rebellion,in S.C. Moitra, *Selections from Jnananneshan*, Kolkata, 1979, p.41.
78. *Ibid.*, p. 40.

79. *Hindoo Patriot,* 19 July 1855, The Sounthal outbreak, in S E PIII, p. 179.
80. *Hindoo Patriot,* 26 July 1855, The Sounthal outbreak, in SEP-III, p. 182.
81. *Hindoo Patriot,* 30 August 1855, The Sounthal outbreak, in SEP-III, p. 189.
82. *Hindoo Patriot,* 21 May 1857, The metropolis, in S E PIV, p. 61.
83. *Hindoo Patriot,* 28 May 1857, Loyal Demonstrations, in SEP-IV, p. 70.
84. *Hindoo Patriot,* 4 June 1857, The Sepoy Mutiny, in SEP-IV, p-79-80
85. Hiren Gohain, *The Idea of Popular Culture in the early nineteenth century Bengal,* Kolkata, 1991, pp. 24-25.
86. *Hindoo Patriot,* 3 December 1857, who are the Rebels, in SEPIV, p. 211.
87. *Hindoo Patriot,* 28 June 1857, Torture by Police in Bengal, in SEP-IV, p. 178.
88. Bankim Chandra Chattopadhyay, *Anandamath*
89. Rabindranath Tagore, *Bichitra,* Kolkata, 1961, pp. 199-205.
90. Tithi Bhattacharya, *Sentinels of Culture,* New Delhi, 2005, pp. 121, 156-157.
91. Anindita Ghosh, An Uncertain Coming of the Book, *Book History,* 6, 2003.
92. Ibid.
93. Priya Joshi, *In Another Country,* New York, 2002, p. 7.
94. Ibid, p. 21.
95. RPI 1878-79, Report of the Inspector of Behar Circle, p. 101.
96. *Note by A. Stirling, 20 June 1824 No.33, in Proceedings of the General Committee of Public Instructions, 14 January 1826, Correspondence and Proceedings, Vol-III, Part-I, Minutes of the Committee, 1823-1838 (WBSA).*
97. *Ibid (WBSA).*
98. Mccleery and Finkelstien, op. cit., p. 116.
99. Terry Eagleton, *Against the Grain,* London, 1986, pp. 13-14.
100. Ibid, pp. 14-15.
101. *The Tempest, op. cit.,* Act I Sc II, p. 364, p. 33, *Othello,* opcit, Act V Sc II, p. 131.
102. Gautam Bhadra, op. cit., p. 5263.
103. Dhirendranath Baskey, *Santhal Ganosangramer Itihas,* Kolkata, 1976, p. 78.
    Ibid., p. 94.
104. Ranajit K. Samaddar, *Bangla Sahityo O Sanskritite Sthaniyo Bidroher Probhab,* Kolkata, 1982, p. 463.
105. Ibid, p. 61.

106. Lili Datta, *Jibonshilpi Manik Bandopadhyay*, Kolkata, 1989, p. 62.
107. *Ibid*, p. 64.
108. English Literature in India, in *The Calcutta Review*, Vol-V, No. IX, January-June 1846.
109. Priya Joshi, *op. cit.*, pp. 229-230.
110. Nanigopal Chakraborty, *Manab Sobhyotar Kahini* (The Story of Human Civilisation), Kolkata, 2003, p. 147, Subhash Ranjan Chakraborty & Goutam Niyogi, *Sobhyotar Itihas* (History of Civilisation), Kolkata, 1998, p. 122. *Note*: These are History textbooks of Class VIII in West Bengal.
111. Provatangshu Maity, *Adhunik Juger Sobhyotar Kahini* (The story of civilization in the modern age), Kolkata, 1998, p. 116, *Note*: Textbook of class VIII in West Bengal.
112. Kalyan Chaudhuri, *Bharater Itibritta* (A History of India), Kolkata, 1987, p. 269, Note: Textbook of Class VIII in West Bengal.
113. R.C. Majumdar (ed.), *The History and Culture of the Indian People*, Vol-I, The Vedic Age, Mumbai, 1988, p. 24.
114. R.C. Majumdar, Glimpses of Bengal in the 19th Century, in Vinay Lal, *The History of History*, New Delhi, 2003, p. 97.
115. R.C. Majumdar (ed.), *The History and Culture of the Indian People*, Vol-X, British Paramountcy and Indian Renaissance, Part-II, Bombay, 1991, p. 1.
116. Vinay Lal, Ibid, p. 96.
117. Priya Joshi, opcit, pp. 229-230.
118. Salman Rushdie and Elizabeth West (ed.), *The Vintage book of Indian Writing 1947-1997*, London, 1997, pp. XIII-XIV.
119. *Ibid*.
120. Salman Rushdie, *Midnights Children*, London, 1995, p. 424.
121. Ibid, p. 194.
122. Ania Loomba and Martin Orkin, *Post-Colonial Shakespeares*, London, 1998, p. 150.
123. Jonathan Dollimore, 'Shakespeare and Theory', in Ania Loomba and Martin Orkin (ed.), *Ibid*, p. 268.
124. *Ibid*, p. 269.
125. Viney Kirpal, *op. cit.*, p. 61.
126. Stuart Blackburn and Vasudha Dalmia (ed.), *India's Literary History*, New Delhi, 2004, pp. 1-2.

Stuart Blackburn and Vasudha Dalmia (ed.), *India's Literary History*, New Delhi, 2004, pp. 1-2.

National Identity and the Ideas of Development in Modern India : Vision of Maulana Abul Kalam Azad

Sujaya Sarkar

The ideas of nation, nationalism and national identity form a basic field of discussion in modern Indian history. Benedict Anderson observes that nationalism requires a conception of an 'imagined community', which he describes as a sense of 'deep horizontal comradeship' stretching far beyond face to face proximity. [1] The term 'nationalism' is generally used to describe two phenomena: a) the attitude that the members of a nation have when they care about their national identity and b) the actions that the members of a nation take for seeking to achieve (or sustain) self determination. [2] It is often defined in terms of common origin, ethnicity or cultural ties. In brief, nationalism emphasizes collective identity a 'people' must be autonomous, united, and express a single national culture. [3] Nationalism, in modern concept, is considered as the key to a holistic as well as sustainable development of individuals, communities and the nation. Since the time of independence, creating of such collective unity received the most priority in government policy formation. Let us observe the significance of the idea of national identity in the formation of a modern nation state and how the idea of development directly relates to this.

In this context, the ideas and role of Maulana Abul Kalam Azad, one of the most renowned national leaders of modern India are analyzed, how he had visualized the importance of national unity in the nation-building process and the total development of our nation. Emphasis is also put upon the objectives and challenges connected with the growing aspects of national identity, as observed by him.

Maulana Abul Kalam Muhiyuddin Azad is regarded as a renowned political leader of Indian freedom movement, a veteran Muslim scholar, a strong opponent of the partition of India on communal lines, the first Education Minister of the government of independent India and a sincere propagator of the ideas of secularism and socialism.

As a matter of fact, the ideas and contributions of Moulana Azad are less discussed for a long time. First scholarly study on Azad appeared long after his death and was even not written in India, but at Oxford. [4] Most of the discussions on Azad's contributions are based on his political activities. The book written by him and the journals edited by him mostly dealt with the political affairs. The main motto of his political activities and writings was to criticize British government and to espouse the causes of Indian nationalism. At the same time, he had to face some great tragedies. He was personally unwilling to accept the principle of the two-nation theory. As the President of the Indian National Congress, Abul Kalam Azad, in 1940, cautioned the country against undermining India's national solidarity. In his presidential address of 1940, he observed, 'Eleven hundred years of common history have enriched India with our common achievements. Our languages, our poetry, our literature, our culture, our art, our dress, our manners and customs, the innumerable happenings of our daily lives, everything bears the stamp of our joint endeavour. There is indeed no aspect of our life which has escaped this stamp'.[5] On the day of partition of India, his dream of a unified nation was shattered. He expressed - 'When I first became aware of such suggestions I was shocked. I immediately saw that this was a dangerous sentiment and could have the most unfortunate and far reaching repercussions. It implied that partition was being accepted on the basis that in both India and Pakistan, there would be hostages who would be held responsible for the security of the minority community in the other state. The idea of retaliation as a method of assuring the rights of minorities seemed to me barbarous. Later events proved how justified my apprehensions were. The river of blood which flowed after partition on both sides of the new Frontier grew out of this sentiment of hostages and retaliation . . . Thus India gained her freedom but lost her unity'.[6]

In Indian history, the twentieth century witnessed some momentous developments. It acquired political independence after prolonged freedom struggle after being remained colony for centuries. With decolonization, India emerged into modern nationhood, where diverse people with myriads of dissimilar religions, languages, ethnicities and regional identities unified in the course of the anti colonial struggles and decided to live together as a unified nation. But that time was crucial in the history of India. From a socio-political perspective, it can be said that the government of India has faced one of the most difficult problems of that time of creating a nation-state from diverse congeries of people, often from a scratch. The other basic problems which were inseparably blended with the building of the structure of the newly created state were the discrimination of gender, minority problems, unsatisfactory standards of human rights and overall inadequate standard of education and literacy. But still we can say that with all those obstacles, the leaders of the nation tried to contribute in the process of structural adjustment of the country then.

In this perspective, Azad, along with his vast political activities, simultaneously made his contribution in shaping various progressive ideas related to the then Indian society. His social awareness prompted him to play some decisive roles. It was to him a life time struggle to lead to the achievement of freedom in true sense. In his writings and speeches, everywhere he aimed towards enterprising a common national feeling in India.

Abul Kalam Azad's prime concern towards re-orientation of the society was based on various aspects. Firstly, his concern was religion; equally he broadened his vision to the society and education the two main pillars of modern societal construction of India. His areas of attention were

1. modification of the ideas related to Islam as a religion as prevalent then,
2. up gradation of Indian Muslim community in accordance with the nationalist ideas of Indian fabric,
3. shatter the discrimination of gender with special thought for the oppressed Muslim women,
4. he extended his thought towards the deprived and unprivileged sections of the society, and
5. raise the standard and pave the right way to spread education among the whole countrymen all these themes were important to form an unique unified Indian nation.

Azad's prime basic concentration was to fight against the orthodox religious ideas, which were prevalent in the Muslim common people, and specially with the attitude of the Muslim religious leaders of that time. Because of his orthodox Sufi family background, Azad was bound to learn and follow conservative, so-called Islamic tradition in his childhood. But later, he also learnt western education, world history and politics through self-study. Simultaneously, he learnt a lot about various schools of theology and exchanged views with the followers of different religions, such as Christian, Jewish, Parsi, Bahai, Hindu and atheist.

His attention was more critical and diverse on examination of religious texts, as well as traditional Islamic precedents. Azad realized, 'Religion itself, instead of creating harmony, was the greatest cause of dispute in the social history of man. There cannot be either multiplicity or contradiction in the truth. Where there is opposition and dispute, there is no truth'. [7] He believed that 'all religions have two aspects, one of which forms their essence, the hard core of their truth. Another aspect is the outer garb in which they are clothed—in the first aspect, that is *Din*, all religions are essentially the same. All the differences between religions relate to the second aspect, which is the *Shariah* or the external texture of religion consisting of laws, customs and modes of worship.' [8]

Such consciousness evoked in him more fundamental philosophical questioning: 'What is the object of life?' [9] He considered religion through his views of 'materialism' and 'rationalism', which inspired him to fight against ignorance and conjecture. He expressed 'Religion aims at the welfare of humanity, but humanity has to pass through different conditions in every age and in every country. Different nations are at different levels of culture and intelligence. Therefore, when religion appeared in these nations, it prescribed for them a different set of laws in accordance with their level of culture and intelligence. Thus Shariah or Minhaj differed in each nation and whatever shape it took was appropriate to the conditions of the time and the level of culture attained by each nation, but Din or the essential truth of religion was the same for all. This is the Quranic stand.'[10] In his perception of reformation, he realized that the main obstacle before the Muslim community for upgradation were the conservative beliefs and customs.

His emphasis on the Quran, as a logical guide in everyday life, renewed significance to the study of the scripture. He spoke in a new language to the Muslim community in his weekly Al Hilal. It was not only a new language in thought and approach, even its texture was different.[11] From Azad's writings, it is clear that he was thinking of a movement for the revival of argument-based personal belief in Islam. We observe Azad's willingness in changing Muslim attitudes, as, 'Enjoining the good and forbidding the evil'[12] which means moderation, or negative of extremes, which he meant as a true balance indicating the exact middle and by extension, path of justice.

According to him, Islam was democratic by nature. He wanted to introduce democracy of Islam religion, specially in the thinking of Muslim leaders. Azad targeted to attack, what he regarded as, the lack of democracy within the Muslim League. According to him, the Quran itself taught democracy and the parliamentary system. He quoted the scripture, where Prophet said, 'In your affairs take counsel with one another'.[13] He critised several times in his writings the ulama for their stagnation and obscurantism. Even to emphasize the ideas of modernity, democracy and equality in Islamic ideas, he compared these with the ideals of French Revolution. He listed five basic principles of the French Revolution and claimed that every one of them had already been present in Islam for centuries: (1) Sovereignty is vested in the people and is not personal or hereditary. (2) All are equal. (3) The president of the country, in Islamic terms the imam or khalifa, is appointed by the people. He has no essential superiority over other citizens. (4) All decisions [of the president] must be made in consultation with able councilors. (5) The treasury is the property of the people. The president may not spend it without authority.[14]

Though there were political causes against the alien rule, the inter-religious co-operation in matters of general welfare and charitable works was a matter of concern to Azad.[15] His firm belief in Hindu-Muslim unity still depicts him as one of the important symbols of communal harmony. He advised as '--- If other communities do not treat you well, you should still treat them well. The greater forgive the faults of the lesser---.'[16] Needless to say, secular means the principles of liberty, equality and neutrality towards all religion.[17] Azad's vision towards secularism was based on an uncompromising critique of ultra-religiosity or bigotry. Inspired from modern western thought and specially from the English industrial and French democratic revolutions, his basic approach to secularism was derived from his insight into the complexity of the Indian social structure. His concept of secularism was the fruit jointly of religious reformation and modern enlightenment in the Indian context. Further, the secular idea gathered momentum from India's striving to evolve a unified national identity out of its multi-religious, multi-ethnic and multi-lingual diversities. He always critised the attitude of aggressive communalism of the Muslim politicians for their emphasis on communal issues before the national interest. He rejected the all India Muslim League's ambiguous agenda and communal separatism. According to him, Hindu Muslim unity is the 'first fundamental task of self government'. His strong comment, 'I would prefer to sacrifice swaraj rather than Hindu Muslim unity. For delay in the attainment of swaraj will be a loss to India alone, but if our unity disappears, it will be a loss to the whole world of humanity'.[18]

The need for freedom and upgradation of women within Islamic tradition was a matter of great concern to Azad. Undoubtedly, it was a bold and daring venture at that time. His wandering in various places of India and West Asia, brought him into contact with societies far less rigid than his own home. Add to all this, his confession was that, for years the restraint of former religious conviction had been missing, and the conclusion is inescapable that the time and circumstances were ripe for the experiences he describes. [19] In a writing, he opined that 'In order to bring about changes in us today, an absolutely fundamental principle is that change take place within our homes, and that our women remind us of that voice which we hear outside our homes, and, unfortunately, forget.' [20] There remained the question of *purdah* system.

Long after his death, a case of Muslim Divorce known as Shah Banu Case raised a lot of hue and cry within the whole of Indian atmosphere in 1980s. Muslim clerics critized the judgement as an attack on Islam. An MP moved a private member's bill in Parliament, seeking to exempt Muslims from the purview of section 125. On opposing this bill, the then Minister of State for Home Affairs, defended the Court's judgement by quoting Maulana Azad. Azad had written that, 'Quran takes occasion to re-emphasize that proper consideration should be shown to the divorced woman in every circumstance'. This call 'was based on the reason that she was comparatively weaker than (a) man and her interests needed to be properly safeguarded'. [21] This is one of the examples of Azad's thinking about the degradation of women in society.

It was Azad, who expressed that education for women was more important than that for men because that could create a new consciousness and a new life in the whole nation. He said, 'I need hardly say that all our educational programmes will ultimately depend upon the proper education of women. If women take to education, more than half of our problems will be solved.' He further pointed out, 'Educated mothers will mean children who can be easily made literate.' His vision was reached towards the proper education for adolescent girls and grown up women even within the limitations of financial and other resources. Due to the conservativeness of the society, Azad recommended co-education system for the girl children between the ages of 6 and 11. His opinion is that, 'For the adolescents, the solution is to set apart certain days in the week for girls and boys. I would suggest that three days in the week may be reserved for girls when the school centres will be entirely at their disposal. For grown up women, we may reserve two days in the week for a similar purpose. I know that this is not a perfect solution but in the exiting conditions I feel that in this way alone we can meet the demand for the necessary social education'[22] for the women.

According to Azad, education is the single most important key for social mobility. He felt that, in formulating educational policy, he could contribute to the future form of the country. As, to him, one of the basic problems was inadequate standard of education and literacy. He opined, 'The welfare, progress and prosperity of any State depend upon the quality of its citizens, and this in turn depends upon the education which moulds their character and shapes their thoughts.'[23] As Education Minister, he contributed in a number of important ventures for Indian education.

Azad's basic concentration regarding educational policy was based upon social responsibility, which he considered as one of the prime objectives of the government. Regarding the importance, significance and opportunity of education for common people, Azad professed the concept of the total development of a 'whole man'. He always mentioned it as an essential motto. He said in his speech, 'Every individual has a right to an education that will enable him to develop his faculties and live a full human life. Such education is the birth right of every citizen'. [24] He emphasized on state's duty as, 'A state cannot claim to have discharged its duty till it has provided for every individual the means to the acquisition of knowledge and self-betterment'. [25]

In his struggle against the barriers of modernity, Azad realized throughout his life that there are plenty of masked hurdles before the common people in acquiring modern education in proper way. He felt that those would also hinder the development process of any nation. So consciously, in his every activity and in his every speech he deliberately emphasized the necessity of propagating communal harmony, religious broad-mindedness and cultural cosmopolitanism with the shaping of education system. He opined, 'Today India is free….She can have any kind of mental mould she pleases. Will it be exclusive…. or will it be inclusive, which has been the characteristic of Indian culture throughout the ages?.... In the advancement of nations there is no greater hindrance than narrow-mindedness. It is our duty to keep ourselves free from this disease in the new era of independence which has just begun. There is no other disease so dangerous for the healthy growth of national life'. [26]

Azad's thinking of cultural relationship with the western countries was based on the need for mutual enrichment. During his tenure, cultural and international activities, which had hardly any place in the education system before independence, came into their own. By establishing the three National Academics (Sahitya Akademi, Lalit Kala Akademi and Sangeet Natak Akademi) and the Indian Council of Cultural Relations and by developing relationship with UNESCO in the cause of peace and international understanding, cultural activities received a new stimulus in the country under Azad's initiative. He spoke in terms of a common cultural heritage and of 'world citizenship'. He observed 'the cult of narrow nationalism' which was according to him one of the greatest obstacles to bring changes in the attitudes of countrymen.[27]

Azad had some specific reservations on 'religious education'. He strongly opined that it would not be wise to import religious dictums officially in government schools. He was generous in acknowledging the part of missionary societies 'in the dissemination of modern education and the development of the modern outlook'.[28] In India, he said, the Serampore Mission laid the foundations of European education. According to him, 'The aim of all religious teaching should be to make men more tolerant and broad-minded.'[29]

To avoid centuries of social neglect and in order to equalize and democratize opportunity, under Azad's leadership, several schemes of scholarship were initiated, of which, the most noticeable one was for scheduled castes, scheduled tribes and other backward classes for which the financial provision was raised from Rs. 300000 to Rs. 20 million during his regime.[30]

Thus, education to Azad was not only a mean to literate mankind, but much more than that. He felt that fundamental education is 'one of the surest ways of securing international peace ---- for the peoples of the world'. He specially emphasized on 'social education', by which he meant 'the inclusion of a lively sense of rights and duties of citizenship, both as individuals and as members of a mighty nation' and 'the production of an educated mind in the masses'.[31] He wanted to include 'understanding of the social conditions of the country, health, education, music, drama, dance, poetry, and instruction in universal ethics, including tolerance and mutual appreciation'.[32]

In brief, it can be said, to Azad, 'human material ... is the basis of our national wealth' and for appropriate nurturing of this valuable wealth, his contribution towards developing education system is multifaceted. He wanted to combine fundamental education with social and value-based education. To make education as the means to create identity, Azad's conception was as, 'A truly liberal and humanitarian education may transform the outlook of the people and set it on the path of progress and prosperity'.[33] He added, 'The welfare, progress and prosperity of any State depend upon the quality of its citizens, and this in turn depends upon the education which moulds their character and shapes their thoughts.',[34] by which he dreamed of 'a new national awakening'.[35]

This paper speaks to raise discussion on such a theme basically for two reasons.

Firstly, it can be mentioned that the ideals followed by Azad and his contributions to the nation are still now very much significant and relevant. Azad, as a great visionary, a nationalist to the core, freedom fighter, scholar and educationist, served throughout his life, the twin causes of freedom and justice. Azad once commented, 'I am proud of being an Indian. I am a part of the indivisible unity that is Indian nationality. I am indispensable to this noble edifice and without me this splendid structure of India is incomplete. I am an essential element, which has gone to build India. I can never surrender this claim'.[36] Azad's claim is not any exaggeration. Out and out, Azad was a nationalist thinker and an ardent modernizer. He was found as 'a strange mixture of medieval scholasticism, eighteenth-century rationalism and the modern outlook',[37] which is even the urge of today's India. The secret of his success as a progressive reformer lay in the fact that he practiced in his life what he preached in his works. He was simultaneously, a man of action and a person of contemplation to some definite ideas.

Secondly, the concepts of liberal and progressive ideas constitute the bedrock of Indian nationalism. They were ultimately incorporated and embodied in the Constitution of the sovereign republic of India. Azad's interpretation of the Indian national identity was based on composite and pluralist cultural traditions.

Azad was often lonely and felt detached from his companions for his too progressive mentality. He has been critized by the orthodox religious Muslims and advocates of Pakistan for his perceived affinity and proximity to Hindus. During and after partition, Azad was condemned for not doing enough for Muslim security and political rights in independent India. Even then, unperturbed, he promoted the struggle for freedom in India as a struggle for a secular republic where all of India's inhabitants were entitled to live, irrespective of religious denomination. Azad claimed, 'Whatever I found, it came out of my own search'.[38]

To conclude, it can be said that, Maulana Azad's basic goal was to preserve, consolidate and strengthen India's unity to build up and protect the nation-state as an instrument of development and social transformation. We have to be vigilant about the newly coming challenges towards our nationalist thinking. New concepts of diversity of opinion are being articulated purely in sectional, communal, sectarian, caste-based, regional or extremist agendas. In short, 'localization of politics' is intruding the premises of national scenario with a serious threat to weaken our national fabric. Arguably, a fear of crisis fostering newer concepts of identity politics is very much evident now. India is now in such disarray that we see agitations that have broken out in different parts of the country in recent times. The largest democracy in the world appears to be hanging from a thread. [39] Today it seems that identity politics is not an isolated phenomenon but follows a global trend. [40] Threats could blight this shared sense of identity. [41] Let us not forget that the essential elements of our national identity are imbued with the values of secularism, egalitarianism, opposition to caste and gender discrimination and support for democratic arrangements, above all the creation of an intelligentsia that would become adequate for the task of nation-building and proper development. No Indian can be empowered by diminishing India. The narrow loyalties of caste and religion must be surpassed. And here comes the necessity of retrospection of Maulana Azad. Azad's sense of values, clarity of judgement, generosity and intellectual versatility prompted him towards nationalist thinking. Azad had only one vision of the Indian nation united and indivisible which, in his judgement, neither fantasy nor could scheming separate and dismantle. Azad's thought of unity and integrity is the basic, prime need of the people, where every hindrance would be cornered in its truest sense. To me, this is one of the most important and relevant aspects to explore the then role of Maulana Abul Kalam Azad, one of the less discussed national personalities of independent India.

Notes :

1. Benedict Anderson, *Imagined Communities: Reflections on the Origins and the Spread of Nationalism*, London, 1983, p. 5-7.
2. K. Nielsen, *Cosmopolitanism, Universalism and Particularism in the age of Nationalism and Multiculturalism*, Philosophical Exchange, 29, 1998, 1999, p. 9.

3. John Hutchinson and Anthony D. Smith (eds). *Nationalism*, OUP, 1994, p. 4 -5.
4. Ian Henderson Douglas, *Abul Kalam Azad An Intellectual and Religious Biography*, OUP, 1988, p.
5. Abul Kalam Azad, '*The communal problem*', in R.V. Ramachandrasekhara Rao (ed), *Indian Unity - A Symposium*, Publications Division, Government of India, New Delhi, 1969, p. 73-75.
6. Abul Kalam Azad, *India Wins Freedom*, 1988, p. 216, 246.
7. Abdur Razzaq Malihabadi, *Zikr e Azad*, Calcutta : Daftar 'Azad Hind', 1960, p257 258; Abdur Razzaq Malihabadi (ed), *Azad ki Kahani Khud Azad ki Zabani, Mak taba e Isha 'at al Quran*, Delhi, 1965, p. 394 395.
8. Shaikh Mubarak Ali (ed), *Tarjuman al Qur'an*, Vol. I, Karachi, 1945, p. 217.
9. Abdur Razzaq Malihabadi (ed), *Azad ki Kahani Khud Azad ki Zabani, Mak taba e Isha 'at al Quran*, Delhi, 1965, p. 396.
10. Shaikh Mubarak Ali (ed), *Tarjuman al Qur'an*, Vol. I, Karachi, 1945, p. 217.
11. Jawaharlal Nehru, *The Discovery of India*, 1947, p. 289.
12. Abul Kalam Azad, *Al - Hilal*, 1 (18), p. 9.
13. Abul Kalam Azad, *Al Hilal*, 1 (2), p. 5- 6.
14. Abul Kalam Azad, *Al Hilal*, 3 (1), p9 11; cited in Ian Henderson Douglas, *op.cit.*, p. 132.
15. Proshanta Nandi, '*Visions of Nationhood and Religiosity among Early Freedom Fighters in India*', in S. L. Sharma and T. K. Oommen (ed), *Nation and National Identity in South Asia*, Orient Longman, 2000, p. 135-149.
16. Abul Kalam Azad, *Al Hilal*, 1 (8), p. 2- 3.
17. Partha Chatterjee, *A Possible India Essays in Political Criticism*, OUP, 1998, p. 241 - 244.
18. Abul Kalam Azad, *Presidential Address*, Special Session of the Indian National Congress, Sept, 1923.
19. Ian Henderson Douglas, Ibid, p. 88.
20. Abul Kalam Azad, '*The Hizbullah and Muslim Women*', 28 May, 1913, *Al Hilal*, 2 (21), p. 367.
21. *Lok Sabha Debates*, 23 August, 1985.
22. Abul Kalam Azad, *Press Conference*, May 31, 1948.
23. Abul Kalam Azad, *Inaugural Address at the UNESCO Seminar on Rural Adult Education*, Mysore, November 2, 1949.
24. Abul Kalam Azad, *Broadcast from AIR*, New Delhi, September 30, 1953.

25. Ibid.
26. Abul Kalam Azad, *Convocation Address at Patna University* on 21 December, 1947.
27. Abul Kalam Azad, *Convocation Address at Patna University*, December 21, 1947; Abul Kalam Azad, *Speech of Welcome at the Second Session of the Indian National Commission for cooperation with UNESCO*, New Delhi, March 24, 1951.
28. Azad, *Press Conference*, February 18, 1947.
29. Ibid.
30. Arsh Malsiani, *Abul Kalam Azad*, Publications Division, Ministry of Information and Broadcasting, Government of India, 1976, p. 88.
31. Abul Kalam Azad, *Press Conference*, May 31, 1948.
32. Ibid.
33. Abul Kalam Azad, *Press Conference*, February 18, 1947.
34. Abul Kalam Azad, *Inaugural Address at the UNESCO Seminar on Rural Adult Education*, Mysore, November 2, 1949.
35. Abul Kalam Azad, *Speech at the Conference of State Education Ministers and Vice Chancellors of the Universities*, New Delhi, April 18, 1953.
36. Abul Kalam Azad, *Presidential address at the fifty third session of the Indian National Congress*, Ramgarh, March, 1940.
37. Jawaharlal Nehru, ibid, p. 290.
38. Abdur Razzaq Malihabadi, *Zikr e Azad*, Calcutta : Daftar 'Azad Hind', 1960, p251; cited in Ian Henderson Douglas, *op.cit.* p. 29.
39. Partha Chatterjee, *op.cit.*, p. 79.
40. Michael Dusche, *Identity Politics in India and Europe*, Sage, 2010, p. 99.
41. Ellen L. Frost, *Asia's New Regionalism*, 2009; Lina M. Kassem, Anthony N. Talbott and Michael T. Snarr, '*Nationalism*', in Michael T. Snarr and D. Neil Snarr (eds), *Introducing Global Issues*, 2010, p. 37-55.

Educational Scenario of the Tea Garden of Duars and the adjacent region in the post-colonial period

Ramendra Nath Bhowmick

The history of educational scenario specially at the tea garden of North Bengal is an issue which may open a new phase if historians, social scientists, sociologists write history regarding development of education among the tea gardens labourers of Duars. This Duars region before 1947 was supposed to call as 'Non-Regulation' area. So, the administrative set up of this region was separated from other parks of West Bengal. For that reason the process of educational development was going on very slowly specially in this region. An astonishing fact is that nobody has chosen this issue as a research articles or nobody has tried to discover the painful history of education for tea garden labourers in Duars. I acknowledge this fact that it is not easier for me to write about this theme from rational and proper historical perspective due to lack of original books, monograph, archieral record on the tea garden. Therefore, I have prepared this research paper based on some secondary sources and I have collected information from experienced age old person and I have also given emphasis on socio-economic survey. Naturally, the relevant questions may come to the forefront of historians, scholars and social scientists i.e. what was the positive role played by tea garden owners for the development of education among labourers? Or what was the role played by Govt. in spreading education among them? Or after finishing their course where will be the distination for stand by? Apart from this, what was the condition of higher education of tea labourers of Duars region? And the last question is provocating one i.e. why they are now demanding for the establishment of Hindi-Medium college in this region? Most of the labourars of Duars tea garden have come from Chotanagpur, Ranchi, Lohardanga, Santal Pargana, Palaomo, Ganjam, Koraput and tribal inhabited area of Madhya Pradesh. When they left their homeland, they had taken 'Sadri-language' as the link language. [1] I do not know whether the tribal students of Duars region are proud of their mother tongue Sadri or not. But true to say that the Bengalees are wholeheartedly proud of about their mother tongue. Now tribal school and college had Bengali medium in school and colleges. Though their mother tongue was not Hindi or Bengali. Why is it so? I have tried my best to discuss the above mentioned questions from rational historical perspective.

I

The first tea garden was opened by Dr. Brougham, an old - hand in tea plantation at Gazoledoba on the left bank of Tista in the western portion of the Duars in 1874. [2] So, at the beginning, education of tea garden labourers was not so developed comparatively now as it is, probably became plantation labour was appointed on a three year contact basis. The plantation labour Act, 1951 made provision for educators facilities for the children of the labourers and some primary schools were opened in the tea gardens. These schools were managed by the District School Board and are now being looked after by the District Inspector of

Schools for primary education. [3] Government also ordered in 22 March, 1916 that school should be constructed for the wards of workers in tea gardens. At first, such schools were under Govt. control, but later they were made over to the tea garden authorities. There were about 350 primary schools in undivided District of Jalpaiguri in 1927-32 of which one was a Government school, 201 were managed by the District Board, 2 were managed by the Municipality and 841 were private primary schools. [4] Sometime christhan Missionaries have contributed a lot in spreading education in tea garden area of Duars. The Roman Catholics started the St. Capitamic Schools for Girls in Nagrakata in 1948, which become a Junior High School in 1958 and a High School in 1963. A co-educational school named Nangdala Missionary High School had founded at Birpara. There were few missionary school for girls located in the Dengnajhar Tea Estate, Malbazar and Nagrakata, one of them being the St. Mary's Boarding School. [5] But it is stated by K. B. Yogi that the tea gardens are private properties and hence it is the duty and responsibilities of the tea garden owners to look after the welfare of their workers under the provision of out dated plantation labour Act of 1951. [6]

All the tea garden except one, surveyed in the Duars have schools which come under the purview of the District School Board, Jalpaiguri. The income of Board is largely provided by a cess of which about one third is contributed by the tea gardens. The Board usually pays Rs. 18 per month to the teacher and this salary varies from Rs. 25 to Rs. 40 per month in different gardens. It was complained by the Indian Tea Planter's Association, Jalpaiguri that the School Board did not sanction the opening of new schools in Tea Eastes. [7] Jodedars formerly objected to their children going to school or being taught, even at home, to read and write. They had an idea that children would desert ploughing and agriculture and take to writing petitions or doing "Babugiri" (acting the Babu). But times are changing; Jodedars are now subscribing out of their own funds for establishing primary schools in the Duars. The amount spent by Jodedars themselves for school during the last three years is Rs. 3,340.[8] W. W. Hunter in the 'A Statistical Account of Bengal' stated that in 1870-71 there were 64 Government and Aided schools in Jalpaiguri, with a total of 1372 pupils, educated at a cost of Rs 855 of which Government contributed Rs. 463. In 1871-72 total number of schools were 35 and number of students 736, educated at a cost of Rs. 758, of which Government contributed Rs. 374.

In 1872-73, there were altogether 61 schools returned in the Education Report (including 10 unaided but inspected schools), attended by a total of 1358 pupils, educated at a cost of Rs 854, of which state contribution amounted to Rs 354. It is evident, therefore, from this staticstiction, that the state of education in Jalpaiguri and specially in Duars tea garden made no apprciable increase during three years and that in 1871-72 there was a very considerable drop, both in the number of schools and pupils, which is not commented upon or explained in the Education Report for that year.[9]

After the independence for improvement of the working conditions of the tea plantation workers, several Acts have been promulgated by the government, such as, The Plantation Labour Act, The Provident Fund Act, The Gratuity Act and the Bonus Act etc.[10] So we see so many Act was passed by government for the development of socio-economic condition of tea labourers of Duars. Now it is apt time for the researchers, historians and social scientists for further study on this issue. Why did the Govt. pass so many Act without Education Act both by state govt. and central govt.? Is it not like that govt. or tea garden owner will provide food and shelter not education? There is enough scope for historians and sociologists to discuss this topic.

The following figures regarding schools and attendance are supplied by the Indian Tea Association.

Table - I

Number of schools and average daily attendance in Duars tea gardens, 1942-43

Year	Number of Schools	Number of Pupils	Average daily attendance
1941	150	6,732	4,218
1942	146	6,723	4,569
1943	149	5,130	3,423

So, we found the fluctuation in statistics on both in case of number of schools and number of students of tea estates due to lack of proper steps from the tea garden owners and the govt. in between 1941-1943. [11] Ranjit Dasgupta in his book "Economy, Society and Politics in Bengal : Jalpaiguri 1869-1968" sharply discussed the deplorable condition of education of Tea labolurers. He had written administrative system of the tea garden, selection of land, recruitment and retrenchment of labourers, transport system, plantation of tree, weekly market for the labourers, formation of District Board in each and every step Govt. played their positive role. [12] In this regard, Govt. activities should be highly appreciated. But for the development of education in Duars tea garden area Govt.'s role was not positive i.e. by establishing new school, selection of proper place for school building, recruitment of teachers and payment of salary, by giving books with free of cost, student's scholarship, by giving school uniform among students etc.

All the historians, social scientists, researchers will acknoledge the fact that English schools sometime have played their prominent role in spreading education among the tea garden labourers of Duars. One Saheb had written to a deputy inspector of the district -

To,

The District Deputy Inspector of Schools
Jalpaiguri

No. 12/1939 dated Baradighi the 25th April, 1920

I am sending a Money Order for the sum of Rs. 36 to your address. This represents the teacher's salary and contigencies for the month of January, February and March 1920.

Please sign and return the enclosed receipt as soon as possible.

From,
The Honorable Mr. W. L. Travers,
O.B.E. Chairman,
Dooars Planters' Association. [13]

In nurschering tea garden - plucking tea leaf, using fertilizer, doing Howai and using Oxin Hormon and pesticides and preparing land in each and every steps there needed experienced and adroit labourers. For the development of tea garden, tea labourer played a very prominent role. [14] It is true that sometime tea labourers - their callousness and lack of consciousness in education was indirectly responsible for slow development in education in Duars and the adjacent areas. Because they have shown their dual opinion in the field of education. Those who are comparatively aged thought that there is no need of further education because they have to work in the garden. One interesting fact is that *Virginia Khaka,* first started his career as tea garden's labour and after completing his higher education, he joined as a professor under North-East India University. [15] It is nothing but the rising ray of hope besides the painful history of education in Duars.

In every tea garden percentage of women education was very poor not only during coloning rule but also after independence this picture has not changed. (average from 16 to 20 percentage). After the introduction of 'Development Cess' owners of tea garden have tried to decrease two type of expense - (one) expenditure on education. (Two) expense on medicine for the treatment of labourers. Because proprietors of tea garden thought that as they are giving 'Development Cess' so full responsibility goes to Govt. for the development of education in he region. [16] Apart from this, there was another great problem i.e. the problem 'Drop out'. If 50 students got admission in class - I then maximum 30 students will remain in class IV. 20 students those who have not attended classes, among them maximum students are female. The number of school going girls were decreasing day by day from primary stage to higher stage. Very few students were able to reach college for their higher studies. [17] One thing I can not but to acknowledge that cent percent tea garden's labourers were women and for this reason here I have started with women first.

II

In every tea garden, there was a common problem. This problem is known to every social scientists, historians, researchers and above all to every educationist as language problem. Now we may find out some relevant fact that how far this problem created an obstacle on the way which ultimately badly affected the then education system in Duars. Social scientists have stated that motherland and mothertongue - two things acquire by human being at the time of birth. But in this regard, prof. Ananda Gopal Ghosh has commented that "The right on mothertongue does not come as the right on land come specially birth time". That is the labour who come of in Duars region, may the son of soil of Duars but he may uses the language of his forefather, migrated from Chhotanagpur plane of Chhatrishgarh region. Because it is their mother language. One thing everybody will acknowledge that nobody can snatch aomeone's mothertounge. Because this language is wholeheartedly related with his or her occupation.[18]

Although after few years, the establishment of tea garden had extended in some parts of Coochbehar and in North Dinajpur, but there was no such language problem as I mentioned earlier. Social scientists have argued thet the inhabitant of this region will never face language problem. Why is it so? The labourer who migrated internally from Madhyapradesh or Chhatanagpur or Chhatrishgarh, his or her mothertoungue was not Hindi. In order to maintain their daily work, they use link language which is called 'Sadri'. But the son and daughter of the tea garden's labourers used to study through Hindi language. After completing secondary education, they are compelled to go to other Hindi Medium State to take higher education. Under this circumstances, one group vehemently oppose the educational policy of state govt. and central govt. and therefore they are now demanding to set up Hindi Medium College in Duars region. But problem has created due to some other reason. Because there is only three state approved language in West Bengal i.e. Bengali, Nepali and Urdu. So, it is not possible for state govt. to open separate Hindi Medium College with state fund in that region. In this regard, we can remind the tea garden's labourers of Assam, they are not demanding for the establishment of separate Hindi Medium College there. So, the relevant question may raise - why did the govt. of West Bengal never think to introduce Bengali as the principal language as medium in secondary and higher secondary level? Because Hindi and Bengali both these languages are not mothertongue of tea garden's labourers of Duars. [19] Sudhir Kumar Bishnu commented that Austric - 19, Dravid - 12, total 31 category of tea labourers can understand Sadri language. Nearly about 60 percent tea labourers though speak in Sadri language and rest 40 percent speak in Nepali language. [20]

Father Franchis Bikka published book named 'Krusaker Rasta' through Debnagari script in 1970. Perhaps it is the first poetic drama written iV Sadri language. Labour leader like Shri Jagannath Oraon published a periodic newspaper called 'Laljhanda' which was highly accomplished by Shree Oraon. After few month, this periodic paper stopped to publish. [21] If cleo meteculously judge it, then it will be clear to us that they were also responsible for it. As we can say during industrial revolution in England, Industrialist had taken huge indentured labourers from India to flourish their industry. Labourers those who have migrated from India to England and other colony, they had griped education to sustain their language, culture, heritage and customs. Till today these labourer are able to maintain their Indian lnguage, social customs, religious rites etc. For example we can say that indenture labourer of Trinidad had given names of some localities as Calcutta Settlement, Madras Settlement, Patna village, Beneras, Fyzabad etc. Road's names as Agra, Ayodha, Bengal, Lacknow avd Madura etc. [22] So, if we seriously study it why did the indentured labourers sucess to established their power in other state? And why did the tea garden labourers of Duars fails to do so? Though the nature and pattern of language problem both in Duars and outside India was same. Here one thing is vivid to other that the labourers of Duars have failed to adopt 'Sadri' as their own language. Apart from this we do not observe any positive steps from both state and central Govt. Because Indian Govt. has not incorporated Sadri language in the 8th schedule of Indian constitution. So the tea garden's labourers and their sons and daughters are complled to study through Bengali and Hindi instead of Sadri and the educated labourers specially those who have acquinted mastery over the Sadri language, they are losing interest to study this language due to the demand of the era.[23]

Though increase of Assam we find to some to some extend different picture. In Assam tea garden's labourer and their sons have been enjoying reservation in Medical and Engineering college. There were some residential hall for students. Apart from this, they were getting books free of cost. In this regard, Assam Tea Employees Welfare Board have played its prominent role.But we do not know that is there any kind of organisation in Duars or not?[24] There was a rationing system in Duars through which rice, dal, wheat etc. distributed among the labourers. In 1858-59 'Indian Tea Planter's Association' Jalpaiguri had taken so many measures for betterment of tea labourers which could be seen in the pamplet of Executive Committee.[25] So, mentality of the gardener reflect that we will provide you rice, dal and wheat but not education.

Whatever may the cause behind the education system of Duars. But one question always hunts us that why are they now egar to establish Hindi Medium college at Duars? Another question - why are the students of Duars happy after getting question through Hindi language? So, for the first question there is needed impartial discussion both from govt. and tea gardener. As there is no Hindi Mediul College, so, after secondary and Higher Secondary education, students of two region are compelled to go other state to have their higher education. So, it is very much difficult for them to complete their graduation from here.

For this reason, it is necessary to establish Hindi college at Duars.[26] Now these question will arise if we establish Hindi college then all problem will solve? or each and every students of Duars will be graduate? So proper place should be identified by thouughtful intellectual or social scientists.

This year some students of 'Red Bank Tea Estate' like Neha Khatun, Tanushree Rai, Sapna Kujur have proclaimed that "We had to face difficulty at the time of Madhyamik Examination due to language problem. Very often we could not understand the inner meaning of the question. But this year we are not facing such problem. We are very very very happy". Headmaster of Banarhat High School states tat "We are also happy as students. Because teachers were compelled to make clear about the question before sttudents. But sometime we could not understand questions due to some problem.[27] Now again some relevant questions comes to our forefront. If questions were set in Hindi language then every students will carry above 80% marks? Or every teacher will understand each and every question's inner meaning? On this issue, there may be some hotbed debate. But everybody will acknowledge the fact that language is a great problem for the expansion of education in Duars. Here history is like pregnant women. Because as nobody can predict that pragnant women will give birth dead child or female child or delinquent child or physically handicap child, similarly history of education of plantation labourers of Duars is absolutely uncertain to us. Because from where it had started and where it will stop, it is not known to us.

Notes :

1. Dr. Sudhir Kumar Bishnu, "Duars - a Sadri Bhasar Utso Sandhan", Duars Samachar, Sarad Sankha, 1416, editor - Ramen De, p. 199.
2. Ranjit Dasgupta, "Economy, Society and Politics in Bengal : Jalpaiguri 1869 - 1947", Oxford University Press, 1992, p. 56.
3. "Gazetteer of India", West Bengal, Jalpaiguri, published by Dr. Barun De, Honorary State Editor, West Bengal District Gazetteer, p. 267.
4. Ibid., p. 258.
5. Interview, Date 07/05/11, Dr. Patit Paban Tarafdar, Age - 75, Athelbari, Jalpaiguri.
6. http://gorkhapedia.wikidot.com/printer friendly//demand for the seperate state of Gorkhala accessed on 11.06.2012.
7. B. C. Ghosh, "The Development of tea industry in the district of Jalpaiguri, 1869-1968, printed by Newman's printer, Calcutta, 1970, p. 57.
8. D.H.E.Sunder, "Survey and Settlement of the Western Duars in the District of Jalpaiguri, printed by The Bengal Secretariate Press, Calcutta, 1895, p.90.
9. W.W.Hunter, "A Statistical Account of Bengal", Concept Publishing Company, New Delhi, 1877, pp.314-315.
10. A.K.Niyogi, "The Life of Plantation Workers in the Past and the Present", Nedam Tea Estate Centenary 1984, Northern Bengal Tea Corporation Limited, Jalpaiguri.
11. B.C. Ghosh, op.cit., p.58.
12. Ranjit Dasgupta, op.cit., p.56.
13. "Detailed Report of the Central Committee of the Duars Planters' Association for the year 1920", printed by Cathilic Orphan Press - 1921, p. 359.
14. Shivaranjan Mishra, "Tea Industry in India", published by S.B.Nangia, Ashish Publishing House, 1986, p.148.
15. "Interview", Dr. Ananda Gopal Ghosh, Prof. North Bengal University, Department of History, Age - 62, Date, 25.03.2011, Time - 4 pm.
16. Manas Dasgupta, "Chabaganer Stri Shramik", 'Samhati' West Bengal Tea Garden Employee's Association, Regd. No. NB/149, Souvenir - 1995, Published by Sankarlal Das.
17. Ibid.

18. Dr. Ananda Gopal Ghosh, "Sabuj Upanibasher Janma bhasa O Janmabhumir Samasya", 'Janamath' Sarad Argha 1416, (Bengali year), Secretary, Mukulesh Sanyal, p. 59
19. Ibid., p.60
20. Ram Abatar Sharma, "Cha Silper Bartaman O Bhabisyat", Publisher by Sudhir Kumar Bhattacharya, First publish, December 2010, p. 97
21. Dr. Samir Chakrabarty, "Jalpaiguri Jelar Sadri Bhasa Charcha", "Kiratbhumi', Jalpaiguri Jela Sankalan, (Ditiyo Khanda), Secretary, Arabinda Kar, pp. 699-700.
22. Dr. Ananda Gopal Ghosh, "Bagicha Chramik Theke Rastra Prodhan : Sagar Parer Bharotiyo Banoadbhutoder Sarkahan", 'Angikar' 1416 (Bengali Year), Sarad Sankha, Secretary, Hiten Nag, pp. 6-7.
23. Ramen De, Op.cit., p.116.
24. Arabinda Kar, op.cit., p.716.
25. "Indian Tea Planter's Association", Jalpaiguri, p.41.
26. "Jalpaigurire Hindi College Kholar Dabi", 'Uttar Banga Sambad', 11 March, 2011.
27. "Hindi Bhasay Prasna Patra peye Khusi Adibashi Purnyara", 'Uttar Banga Sambad' 3 March, 2011.

ARMY AND THE COLONIAL LEGACY

The Regiment as an Institution : A Colonial Relic?

Sabyasachi Dasgupta

Introduction

The Regiment is a hallowed institution in the Indian army .Traditions, customs, values etc are built around the institution of the regiment. It is almost as if the post independence Indian army is a conglomeration of self contained units who do not necessarily constitute a whole. It would not be incorrect to say that the basic ethos of the army is constructed around the regiment as a unit. To put it bluntly the regiment seems to be the basic functional unit of the army. The logical corollary which therefore seems to follow is that regimental traditions acquire a sacred hue. Every regiment develops distinct customs, traditions and guards them fiercely. Regimental centre mess rooms, colours,battle honours, silverware, trophies presented to the regiment etc are treated as an object of reverence.

Now the question is that when, how and why did regiments become such hallowed institutions? The origin of most of the current Indian and Pakistani regiments are derived from regiments of the colonial Indian army. The colonial army in a manner akin to that of its mother army,i.e the British army attempted to construct a European style army based on regiments and battalions. Or rather it till 1895 sought to build three distinct armies namely the Bengal, Madras and

Bombay armies. While the ethnic composition of these armies was vastly different there were some basic similarities as far as the building blocks of each army was concerned i.e, the armies were centred around the institution of the regiment.

There were certain concrete reasons why the colonial army sought to emphasise on the regiment as an institution .The colonial army was mercenary in nature. The sepoys could not be expected to exhibit patriotic sentiments towards the British cause. At the same time the colonial army was uncomfortable with the notion of the sepoy as a purely mercenary soldier. It therefore strove to build certain indices of loyalty .Promoting a paternal style of leadership where the sepoy looked upon the European commanding officer as a father figure was one of the many strategies they adopted to secure the loyalty of the Indian recruit.

While paternalism as a philosophy of command has been discussed in detail elsewhere we now come to the central focus of our discussion, the regiment. The regiment occupied a central place in the overall colonial strategy of securing the loyalty of the sepoy. The reasoning was that if it was impractical to expect the sepoy to show loyalty towards the crown it was perfectly reasonable to hope that the sepoy develops a loyalty towards his regiment. It was not too much to imagine a scenario where the sepoy would fight to the death for the honour of his regiment, where he would do his utmost to uphold the proud traditions of his regiment. [1]

'The Colonial legacy of the Regiments'

While the three armies followed a common policy of inculcating regimental loyalty, the process spelt itself out in different ways at the ground level at the various presidencies. This was partly due to the varying recruitment policies the presidency armies followed.

To cite the Bengal army's recruitment policy, the Bengal Army was recruited mainly from Oudh and Bihar. High-caste recruits from these areas were supposed to be loyal, obedient, faithful and ideal soldier material.[2] Sleeman, one of the votaries of the Bengal school of recruitment praised the deference of the Bengal sepoy for constituted authority. According to him, in the families from which the Bengal Army drew its recruits, it was unthinkable for a son to utter a disrespectful word to his parents. He comments: "Such a word from a son to his parents would shock the feeling of the whole community in which the family resides and the offending member would be vested with their highest indignation". [3] In the eyes of the Company officers, high-caste sepoys were a morally and physically finer race in comparison with troops of other presidencies. [4]

As a result of this high-caste preponderance, caste became a major factor in the functioning of the Bengal Army. Convinced of the immutability of caste sentiments among its predominantly high caste recruits, caste sentiments and prejudices came to be tolerated and on occasions even encouraged in the Bengal Army. For example, the Bengal sepoy refused to strike the gong at the quarter guard fearing a loss of status. Men called *Ghante Pandeys* were recruited and paid for the job.[5] Taking into consideration high-caste sentiments regarding sea voyages, overseas service was voluntary until 1856, when a general order made overseas service within or beyond the territories of the Company compulsory.[6]

The general mode of recruitment was to ask sepoys and Indian officers, leaving home on furlough, to bring back recruits from their village.[7] In exceptional circumstances, recruiting parties were sent to Bihar and Oudh. [8] A closed-shop system originated, where ties of clan, caste and residence ensured inter-relationship among the sepoys and Indian officers.

In contrast the Bombay Army, right from its inception, was a mixed one consisting of Marathas, Jews, Konkani and high-caste Hindus from the Bengal Presidency. The infantrymen were in physical build, similar to the Madras Army sepoy. The average height was 5feet, 5 inches, but they were robust and capable of enduring enormous fatigue. The native soldier from Bombay was considered to be cheerful under privations and difficulties.

By the 1820s however, nearly fifty percent of the Bombay Presidency troops were composed of high-caste Hindustani men from the Bengal Presidency. They were recruited in large numbers in the aftermath of the third Anglo-Maratha War of 1817-18 and the Burmese War of 1824-25. Prior to that, the high-caste Hindus did not exceed four thousand in number in the Bombay Army. By the 1830s, the Hindustani soldiers comprised slightly above 50 percent of the Bombay Army.[9]

Experienced officers of the Bombay Army viewed the significant increase of Hindustani soldiers over the local recruits from the Bombay Presidency with concern. This was also contrary to the wishes of the court of Directors of the East India Company, who in 1824 had urged that the armies of the three Presidencies be recruited, to whatever extent from their respective home territories. As John Malcolm put it in his reply to the Select Committee of 1831-32, it was a policy decision of the Bombay Army officers to have a mixture of castes in their regiments. The officers preferred to have not more than two hundred Hindustani men per battalion.[10] Despite reservations regarding the high number of Hindustani soldiers in the Bombay army, they continued to have a strong presence in it. They constituted almost fifty percent of its strength in the 1850s. Thus, the Army mainly constituted of Hindustani and Konkani soldiers.[11]

The Madras army followed a recruitment policy which was superficially akin to the Bombay army though the human material in the Madras army was significantly distinct from the Bengal and Bombay armies. Their recruitment policy though had a high-caste bias in its early days, the recruitment was confined to certain specific groups. For example, the military regulations of 1765 limited recruitment to the Rajputs, Muslims and three unspecified Telegu castes. It was argued that military prowess was limited within certain specific groups in society.[12] The Madras army sought to enlist *purbaiyas* who were seen as possessing a fine physique and a handsome appearance. In 1795, Col. Floyd of the Madras army declared that good recruits could not be obtained from the area south of the Coleroon, since this region produced men of small built, i.e. under the prescribed minimum height of 5' 5''.[13] However, the Madras Army was, by the first decade of the 19th century, composed mainly of low-caste Hindu and Muslim sepoys.[14] Douglas Peers ascribes this preference for low-caste soldiers to the fact that the military service was not the preserve of high-caste communities in the South. As a result they were forced to seek their sepoys from a range of communities.[15] Kaushik Roy adds another strand to the debate by asserting there was a lobby that pursued an open-door policy as regards employment as opposed to the selective enlistment lobby. Roy suggests that in the pre-1857 Madras Army the open door policy won against the selective enlistment school due to certain factors. Since the Bengal army had the first choice of recruits in Oudh and Bihar, it monopolized the better variety of *purbaiya* recruits leaving the second-grade resources for the Madras army.[16]

The present study argues that the induction of low-caste recruits in the Madras army was not entirely due to the non-availability of high-caste recruits. There were deeper ideological underpinnings that guided this choice. It may be suggested that the decision to recruit low-caste soldiers was intended to isolate the native army from local attachments and sensibilities. General Harris, one of the main proponents of low-caste recruitments, listed the lack of religious prejudices and local attachment among other qualities that the low-caste recruits possessed.[17] The objective clearly was to foster a corporate identity with the belief that an army having a corporate identity would be a disciplined and professional one. By the 1790s, the Company's army was primarily composed, according to James Welsh, of five groups. The Mussulman was the first of these. According to Welsh, one-third of the army was composed of Muslims who constituted mainly of Syeds, Sheikh, Mogul and Pathans. He goes on to say that, the Muslim recruits were generally brave, intelligent and enterprising, by and large free from religious prejudices. The strong Muslim presence became a conspicuous feature of the Madras army. The second class Welsh describes was the Rajput, a class according to him not represented in great numbers in the Madras army. He opines that the Rajputs made brave and devoted soldiers if their religious prejudices were honored. The third class that he lists was the Telinga or Gentoo, a race of Hindus who, according to Welsh, were remarkable for mildness of disposition and personal cleanliness, but slacking in enterprise and intelligence.[18] The fourth class was the Tamil or the Malabar caste and the fifth was the class of untouchables and the pariahs, who were few in number.

Welsh's account would lead us to believe that the Madras Army, by the 1790s, consisted mainly of middle and lower castes and a substantial number of Muslims. Facts given in certain other accounts corroborates this argument. Like, the work by H.H Dodwell which provide us with useful statistical inputs as regards recruitment in the Madras army for instance in the late 18th and early decades of 19th century. In 1796, a particular battalion had 126 Muslim recruits out of 249 that is nearly 55 percent. In 1805, there were 151 Muslims out of a total of 272, again a percentage of nearly 55. While this may not have been necessarily the standard rate of intake, it could be safely assumed that Muslims were recruited in large numbers in the Madras Army along with middle Hindu castes while high-caste and low-caste recruits formed a lower proportion of it. [19] This trend, so far as recruitment for the Madras Army was concerned, would possibly hold firm till 1857. A discharge sheet for the Madras Army in 1847 seems to indicate such trends, though the critical question remains, could this discharge sheet be taken as an indicator of general recruitment patterns in the Madras army? [20] Out of eighteen men discharged, nine were Muslims; two were upper caste, one Anglo Indian, while the rest were middle castes. [21]

Thus we see some patterns. The Bombay and Madras evolved mixed regiments in a manner akin to the increasingly mixed regiments we have in the modern Indian army. While the mixed composition of the regiments ensured that regimental traditions would not develop around ethnic, caste or religious factors, the regiments nevertheless developed some distinctive regimental traditions centred around regimental colours,battle honours etc. To give an example the Bombay native infantry was aggrieved because their colours were changed from purple to green. Purple for them signified the colour of royalty. While these mixed regiments were gradually phased out after 1857, the ethos surrounding these mixed regiments would have significance for the distant future when most of the regiments of the Independent Indian army would be mixed. For it showed that regimental traditions could be built independent of indices such as caste, clan, religion and other factors of their like. There were other abstract and real factors around which a regimental identity or regimental pride could be constructed. On the contrary regimental pride could be constructed around the fact that caste dynamics did not play a part in the functioning of the regiment. The very lack of these factors could be to build up a regimental identity.

The ethos of the pre 1857 Bengal army was therefore centered on common ties of caste, clan and religion possessed by the recruits. In most cases, as we have seen recruits often hailed from the same villages. This led to a situation where regimental as well as caste, village, clan pride etc got enmeshed and used to become indistinguishable from each other. This would have significant connotations for the future. Though the old Bengal army was largely decimated as a result of the great war of independence, its recruitment patterns would ironically replicated increasingly from the 1880's onwards. One is obviously referring to the martial race discourse which took centrestage in the colonial army around this time.

Under the martial race theory certain communities predominated in a grossly disproportionate manner to their population percentage. Thus regimental traditions would often be indistinguishable from ethnic or religious traditions in single class regiments. It was almost as if the old Bengal army was reborn under a different garb. The old recruiting material might have diminished drastically but the philosophy operating behind the recruitment policy of the old Bengal army seems to have endured. It was like old wine in a new bottle. And this was to have significant connotations for the regimental ethos of the colonial Indian army and the independent Indian army.

The Onset of Independence

Partition resulted in division of the colonial Indian army, though India got the bulk of the armed forces. The newly independent Indian government declared that recruitment would be open to all Indian citizens irrespective of caste,colour or creed. The rider attached was that regiments more than 100 years old would not be tampered with. That effectively limited the innovations the army could do with regiments as per as recruitment was concerned. [22]

While we need not go again into the nitty gritties of the recruitment policy it would suffice to say a strong regimental spirit pervaded through the regiments irrespective of the nature of the regiments,be they be fixed or same class.The only difference would be a single class regiment or fixed class regiments would build regimental identity in a manner akin to the old Bengal army .Or more recently these regiments would be the inheritors of the same class regiments which sprang out of the ethos of the martial race theory. [23]

Irrespective of the nature of the regiment there were some secular indices which developed regimental spirit. For instance every regiment would possess some battle honours.Often these battle honours would involve successes against Indian powers in colonial times acting as troops of the British. The regiments are so enamored of their battle honours that they seem to be oblivious of the politically incorrect step of celebrating successes against indigenous powers resisting the British colonial onslaught. It seems the glory and the pride the regiment derives from these battles, the sense of identity they seem to seek from these battles outweigh any sense of discomfort over the problem.

To cite some examples the regimental history of the Grenadiers mention Mysore and Seringapatnam as battle honours. Similar is the case with the 5thBattalion, The Maratha Light Infantry.[24]

Army officers interviewed were vehement in their defence of the Regimental system. Squadron Leader seemed to think that the regimental system has served the forces well. Sure there are some problems but again the regimental system has taken care of problems which even the armies of developed countries face difficulty in overcoming .Chinna feels that the emphasis on the Regimental system might cause difficulties if the armed forces have to fight huge battles. However RTS Chinna felt that the regimental system was appropriate for the kind of operation milieu the Indian army will function in the future.[25]

Other armed forces officers interviewed felt the same and their view is that the regimental system was the heart and soul of the army and there was no need to tamper with the system. Major General K.K Ganguly felt that the regiment served as a very good indoctrinating unit which absorbed new recruits. A recruit might not know everything about his country but he will know everything about his regiment or battalion. In fact, it was not only the Jawan but also his family which held in awe the regiment. General Ganguly narrates an incident where a jawan had run away from the face of battle. His family refused to accept him saying that he had let down the regiment.[26]

However, there are dissenting voices which believe that the regimental system has some drawbacks. Major General Gagandeep Bakshi felt that excessive regimental loyalty was detrimental to the well being of the army. The primary reason for a jawan's fighting should be patriotism and not loyalty towards the regiment. Regimental spirit at the cost of love for one's country does not seems good. Overemphasis on the regiment could deter coordination in big battles. General Bakshi was of the view that militant nationalism should be the primary factor which will motivate the Jawans.They should be first prepared to die for the country; the regiment should ideally come later. General Bakshi however rued the fact that this was often not the case. The entire commitment, ethos of the Jawan was centred around the institution of the regiment. This, General Bakshi felt could have serious long-term consequences.[27]

Bakshi's voice represents a dissenting one and in many senses an exceptional one. Most army officers overwhelmingly supports for the regiment as an institution. Major General Carduzo felt that regiments had over a period of time acquired an individuality and personality of its own and the men of respective battalions and regiments would be linked by personal ties. No sacrifice was too great for the regiment .It did not matter to a soldier if he died as long the honour of his regiment was preserved. Regiments, as Carduzo commented often assumed an identity equivalent to religion and in many cases generation after generation of a family would serve in the same battalion of a regiment. It was somewhat of a tribal affair. Often family,clan,village, battle and regimental bonds used to become enmeshed. A soldier whose conduct was praiseworthy could hold his head high not only in the regiment but also in the village.[28]

Carduzo's views were echoed by one of the legendary military heroes India has produced, namely Sam Manekshaw.He was firmly against the reorganization of the Indian army into mixed units on the basis of state population. As Manekshaw comments:

> Are the past glories and achievements of regiments like the Marathas,the Sikhs,the Dogras,the Rajputs,the Madras-----to be forgotten? Are their famous rallying slogans to be done away with? In the name of God, desist from such folly. . . If this

imprudent proposed political decision is accepted by sycophantic generals, forecast doom and calamity.[29]

The moot question arises as to why is the regimental system valued so much? Is it just a colonial legacy? Or does it serve some critical needs of the Indian army? Equally does the emphasis on regimental bonding retard coordination at the division or brigade level? If it is so then why has army continued to emphasise on the regiment as an institution? Is it just the weight of legacy which the army is loath to cast aside? Or is it the typically bureaucratic tendency of continuing with tried and tested system unless it is on the verge of collapse? These are some questions one wishes to pose.

At least one historian of the Indian Army,Stephen Peter Rosen has argued that the Indian army's over reliance on regimental spirit has hindered the army's performance in big battles. The entire outlook is geared towards fighting battles at the regimental level. Rosen refers to certain divisional and brigade level battles in the 1962 war where the Indian army acquainted itself poorly. Rosen contends that Indian single class units speaking different languages and had problems in quickly learning to work together resulted in a palpable lack of coordination among the armed forces.[30]

While it is possible that the excessive reliance on regimental bonds might hinder the Indian army's performance in big battles the regimental system obviously performs certain important functions which more than make up for its shortcomings. While some have been stated there are some which are not stated. One such task possibly is the integration of recruits from insurgency prone areas, namely Nagaland and Kashmir. After all these are areas where the emotional integration of the populace with the state of Indiais admittedly thin. It will possibly be impractical to expect recruits from such areas to respond to calls for patriotism. In such a scenario it makes sense to integrate such recruits into the army through the medium of the Regiment. What stops such a recruit from owing allegiance to his regiment and taking pride in the traditions of his regiment even if he does not feel for the nation?

Certain army officers when asked about such a possibility were evasive and defensive. However General Ganguly observed that Naga and Kashmiri recruits get indoctrinated into the ethos of the Indian army during training. They no longer view themselves as Nagas or Kahmiris but as proud recruits of the Indian army. The various training centers ensure this. Once a cadet passes through the grind, General Ganguly emphasises that he is imbibed in the traditions of the Indian army.[31]

General Ganguly also argues that it is not that the Nagas are not patriotic. Their grouse is not against the nation. After all which government, he says, refuses fund and aid to a region? The problem is with the local politicians and officials who appropriate funds. This grouse against local officials takes the form of insurgency and rebellion against the Indian state. There is no problem otherwise in Nagas joining the army. They view it as an employment avenue in an area where employment opportunities are scarce. General Ganguly says that he has himself organized recruitment melas in Nagaland and the response was immense.

There were other interesting views on the integration of Naga recruits into the Indian Army. Lt Gen S.K. Sharma says that the tribal cultures value bravery, killing the enemy etc. This are consistent with soldierly values. Consequently the Naga finds no problem in adjusting to the ethos of the Army. As part of the process of adjustment Regiment is substituted for tribe. Nagas, General Sharma reminds us performed with exceptional courage at Kargil.Therefore the recruitment of Nagas is not a problem despite the continuance of insurgency in Nagaland.[32]

Major General Bakshi says that the naga recruits are always bunched in mixed regiments. For instance Naga battalions are always clubbed together with the Kumaonis. Bakshi is obviously implying that their fidelity is not to be totally trusted and they need to be kept under surveillance. Conversely he also mentions that the Nagas are exceptional fighters and are possessed with a great sense of humour.He remembers the Nagas nostalgically from the time he had served with them in Kashmir. In a lighter vein he says that the fearsome reputation of the Nagas and rumours to the effect that they were cannibals helped in maintaining law and order in the area.[33]

Conclusion:

To put it in a nutshell the Regimental system has multiple usages. It is a tried and tested system and despite its pitfalls has served the army well. The army is therefore loath to drastically change or tamper with the regimental system. Periodically objections have been raised against the overreliance on regimental spirit, prompted by concerns that it promotes sectarian feelings in the army. Armed forces personnel often show deep-rooted identification with the regiment rather than with the army as a whole. There are fears about the Indian army being found wanting in battles involving the coordination of multiple regiments.

A lobby within the army therefore has pushed for an integrated system of command and coordinated warfare. Accordingly the Centre for Joint warfare Studies was set up a few years ago. This centre promotes research towards improving the coordination between different arms of the armed forces in planning strategies and actual deployment of troops in battle. This would obviously imply lesser reliance on the regiment as practically an autonomous unit in battle and instead push for an overarching etho which provides for large-scale coordination among units. While these need not necessarily mean the total dilution of the Regiment as the basic functioning unit of the army, it nevertheless provides for a lesser role to the regiment.

Despite the presence of what one could say an anti lobby the regiment as an institution has endured. There have been the invariable changes in regimental traditions but the basic structure has survived. The changes have come in the menu being served in messes. The staple food on offer in messes has become increasingly Indianised though Chinese or continental will be served once a week and on special occasions. The day to day food and customs for officers is totally Indianised .When we talk of customs, we could refer to certain customs such as toasting the Commander-in chief being given an Indian touch. Toasts to the President of India are now given in water since it is considered disrespectful in Indian culture to a toast being offered to a dignitary using alcohol. There are of course several other customs which have undergone a subtle change and have been progressively Indianised.[34]

So far as Indianisation is concerned many of the tunes played by the Army band during the Beating retreat have been Indianised. Some old popular British tunes are still played but the accent is on Indian tunes. While this may seem to be a trivial change it is significant in the overall context of a culture of Indianisation of regimental traditions, though some would contest this concept of Indianisation of regimental traditions and would argue that the core regimental traditions in colonial times were not really colonial, they were indigenous in origin. Therefore we should not be misled by the apparent Indianisation of external paraphernalia and say that colonial traditions have been Indianised.

Lt.Gen S.K Sharma claims for instance that the so called colonial legacy was developed from past martial traditions in India where literally everybody fought everybody. [35] Sharma is not alone in thinking along this lines. Kaushik Roy in his illuminating piece on the Martial Race theory tries to delve into the indigenous stands which went into the making of the Martial Race theory. The Martial Race theory as has been argued before constructed certain distinctive regiments with a peculiar etho of their own. One could therefore make a case about the colonial regimental traditions being not so colonial. What could be possibly argued more plausibly is that the colonial government gave its own touch to existing indigenous traditions. [36]

A stronger case for colonial government picking up existing indigenous traditions and leaving their own imprint on them can be made in the case of the old pre 1857 Bengal army regiments. The Bengal army interpreted Hindu upper caste traditions and sentiments of present day Eastern UP and Bihar in its own way and sought to construct an etho for the infantry regiments of the Bengal army which sought to religious traditions and sentiments of the predominantly upper caste sepoys of the Bengal army as interpreted by the colonial army. [37]

Theefore the concept of a colonial legacy being inherited by the post independence Indian army in terms of regimental traditions, customs and ethos from the colonial army need to be seriously questioned .One need not be taken in external symbolism .More basic and deeper questions have to be asked. How colonial were colonial traditions really especially with regard to the regiment?

Irrespective of such debates, one could safely say that the post independence Indian army did inherit a strong regimental etho, the controversy surrounding the origin of the ethos notwithstanding. In that sense the post independence Indian army did inherit a colonial legacy. It was understandable that the colonial government wanted to construct the image of the loyal sepoy by binding him to his regiment. The reasons may be more complex when the post independence Indian army seeks to follow the same path. What are the reasons behind this? The answers may be complex. Possibly one of the reasons is that even the post independence Indian army finds it easier to bind the Jawan to more concrete indices rather than the abstract construct of patriotism or nationalism. The process becomes more complex when the army has to deal with recruits who hail from areas which are loosely integrated emotionally with mainstream Indian nationalism.

Notes :

1. For a detailed discussion on paternalism see Chapter 3 of Sabyasachi Dasgupta's unpublished PhD thesis, In Defence of Honour and Justice : Sepoy Rebellions In The 19th Century, JNU, New Delhi, 2005.
2. Report of the Enquiry Committee of the Barrackpore Mutiny, 1824, Military (Misc.), Vol-11, p. 479, NAI.
3. *Ibid.*
4. Russell Select Committee for Indian Affairs, 1831-32, Minutes of Evidence Vol. 13, p. 567, West Bengal State Archives, Kolkata.
5. John Jacob, *A Few Remarks on the Bengal Army and Furlough Regulations with a view to their Improvement*, London, 1855, p. 38.
6. Field Marshall Roberts, *Forty-One Years in India* Vol. 1, London, 1897 p. 256.
7. One important break the Company achieved as far as recruiting was concerned was that the East India Company would stop the practice of recruiting matchlock men under their own jobber-commander or jemedar who would join directly as an officer, a practice prevalent in Indian armies. Henceforth, Indian officers would rise from the ranks, joining service in the capacity of a

sepoy. Indian officers would, however, continue to play an important role in recruitment. See Dirk Kolff, *Naukar, Rajpurt and Sepoy: the Ethno-history of the Military Labour Market in Hindustan, 1350-1850*, Cambridge, 1990, p. 178.

8. Major General Sir Jasper Nichols, Select Committee for Indian affairs, 1831-32, Minute of Evidence, Vol. 13, p. 113, West Bengal State Archives, Kolkata.
9. Reply of Major General John Malcolm to the Select Committee for Indian affairs, 1831-1832, vol. 11, Appendix B, 4127, West Bengal State Archives, Kolkata.
10. *Ibid*, Appendix B, 4127.
11. *Ibid*, Appendix B, 4127.
12. H. H Dodwell, *Sepoy Recruitment in the Old Madras Army*, Calcutta, 1922, p. 14.
13. Koushik Roy, Recruitment Doctrines of the Colonial Indian Army, 1859-1939. *Indian Economic and Social History Review*, Vol. 34, No. 3, July 1997, pp. 345-346.
14. Henry Russel, Select Committee for Indian Affairs, 1831-32, vol. 13, Minutes of Evidence p. 67, West Bengal State Archives, Kolkata. The Company armies were primarily infantry armies and cavalry regiments were fewer in number. The Madras army was no exception. Therefore the discussion on the recruitment policy of the Madras and the other two Company armies would be a discussion on the recruitment policy of the infantry regiments of the Company armies. Muslims dominated the Madras cavalry regiments.
15. Douglas Peers, *The Habitual Nobility of Being; British Officers and the Social Construction of the Bengal Army in the early 19th century, Modern Asian Studies*, 1991, 25/3, pp. 551-552.
16. *Kaushik Roy, Recruitment Doctrine of the Colonial Indian Army, 1859-1939, IESHR*, vol. 34, No. 03, July 1997, pp. 345-346.
17. W. J. Wilson, *History of the Madras Army*, London, Vol. 3, p.151.
18. Colonel James Welsh, *Military Reminiscences Extracted From A Journal Of nearly Forty Years Active Service In the East Indies*, London, 1830.
19. H. H Dodwell, Sepoy *Recruitment in the Old Madras Army*, Madras, 1912, p.42.In spite of an ideological shift in the recruitment policy, official injunctions against the recruitment of certain so called inferior castes continued till the early decades of the 19th century. The Madras army regulations of 1806 prohibited recruitment from certain so called inferior castes such as Palas, Pariar and Chackliar. See Madras Army Regulations, 1806, Madras.

20. General Orders, Commander-in-Chief hereafter to be referred as GOCC, Madras Presidency, 12th May 1848, TNA.
21. *Ibid.*
22. A.L Venkateswaran, *Defense Organization in India*, New Delhi, Publications Dvision, 1969, p. 188.
23. Kaushik Roy, Recruitment doctrines of the colonial Indian Army, *IESHR*, vol. 34, no. 3, 1997, p. 331.
24. See Major General (Retd) K.D Souza. 'A Royal Tribute : A History of the 5th Battalion, The Maratha Light Infantry, 1800-2005. Also see Right Of The Line: The Grenadiers Association, 1995, Grenadier Association, 1995.
25. Interview with Squadron Leader(Retd) RTS Chinna on the 25th of July, 2008 at United Service Institution, New Delhi henceforth to be known as the USI.
26. Interview with Major General (Retd) K.K Ganguli on the 28th of December, 2006 at his residence, Kolkata.
27. Interview with Major General G.D Bakshi (Retd) on the 24th of October, 2008 at USI.
28. See, Major General(Retd)Ian Carduzo (ed) '*The Indian Army: A Brief History*',Centre for Armed Forces Historical Research, 2005, p. 256.
29. Ibid.
30. See Stephen Peter Rosen, *India And Its Armies*, Cornell University, 1996, pp. 242-243.
31. Interview with Major General(Retd) K.K Ganguli on the 28th of December, 2006, at his residence, Kolkata.
32. Interview with LT. General S.N Sharma(Retd), 25th February, 2006, New Delhi.
33. Interview with Major General Gagandeep Bakshi on 24th October, 2008 at USI.
34. Interview with Squadron Leader(Retd) RTS Chinna on the 25th of July, 2008 at the USI.
35. Interview with LT. General S.N Sharma (Retd), 25th February, 2006 at USI.
36. Kaushik Roy,Recruitment doctrines of the colonial Indian Army, *IESHR*, vol. 34, no. 3, 1997, p. 331.
37. For more details see Chapter 4 of Sabyasachi Dasgupta's unpublished thesis, In Defence of Honour and Justice : Sepoy Rebellions in the 19th Century, JNU, New Delhi, 2005.

The Role of the Paraiyans in the Madras Presidency Army C.1801-1894

Manas Dutta

Lower caste group or rather the marginalised groups was one of the important pillars of the Indian army under the colonial rule. It played a significant role to maintain the colonial hegemony during the British rule in the sub-continent. [1] Very little work has been done on the marginalized or untouchable or labourer castes role in the Army. These gropes played a very important role in the Army basically in lower rank as sepoy, guards, watchman and so on which was overlooked by the military historians like kaushik Roy, Dewitt C. Ellinwood and Cynthia H. Enloe, Douglas M. Peers, David Omissi, Zakaria Haji Ahmed, Lanny Bruce Fields, William R. Heaton, Carol Hills, Henry Russel, Henry Dodwell, W. J. Wilson. The East India company has given new opportunity to the lower caste which gave the occupational social mobility and new socio-economic and cultural strength. Recently, Raj Sekhar Basu's monograph has explored the role of the Paraiyans [2] in the Madras Presidency army under the colonial rule. It also, described the rise and decline of the Paraiyans in the army higherarchy and subsequent changing attitudes of the British military officials as well as the colonial government about them. This article tries to explicate in greater details of the Paraiyan caste's involvement in the colonial Madras army and how did they provide the valuable service in the army from 1801 to 1894 in the subcontinent.

Rarely do we recognise the importance of armies and warfare in shaping the course of modern Indian history. The Army in India, which included the Sepoy Army commanded by the British officers as well as the British units stationed in the subcontinent, had a dominant presence in the colonial era. Till the 1890s, the Sepoy Army included the Bombay Army, Bengal Army, Madras Army and the Punjab Frontier Force. The Sepoy Army or British-Indian Army offered the largest government employment to the colonized Indians. Annually, the Sepoy Army, or Colonial Army enlisted about 15,000 Indians in peacetime. Indeed, the colonial army was the biggest item of expenditure for the colonial regime. About 30 percent of the British-Indian Empire's revenue went to feed the military establishment in the subcontinent. Military service moulded the structure and ethos of colonial Indian society.[3]

Compared to ancient and medieval Indian history, the database for colonial India's military history appears more solid. This is due to the book keeping activities of the colonial bureaucracy, as well as a host of unofficial monographs and articles written by the colonial military officers recording their glorious activities in an exotic environment. The memoirs and journals of the British officers as well as soldiers simply signified their activities in the sub-continent and these kinds of writings become popular both in India and Britain.[4]

However, army played an important role in the lives. It gives immense source of social occupational mobility. Army as an occupation acquired an immense honourable place in the society. Military establishments have important social implications, particularly in developing societies. In attempting to determine the relationship between the army and society, several points must be considered. Social attitudes to military service were important. It regarded as an honourable profession.[5]

The relationship between citizenship and military service is significant. This may range from democratic egalitarianism in which full citizenship included at least the possibility of military service, to pure expediency and efficiency, in which only the best qualified and most willing are called upon. The function of the military as an agency of social occupational mobility may be considered in two ways, does or should the structure of the military establishments replicate the social stratification pattern of society, and does it or should it attempt to alter these patterns?[6]

The Occupational Caste groups:

Writings on caste very often fail to ignore the popular conceptions associated with this phenomenon. Caste as a concept speculates interest among the sociologists and historians around the world. Caste played a very significant role in Indian society and certainly it also gave an impact on Indian army.[7] In India, the caste system derives its ideological sustenance from sacred Hindu texts, which emphasised on a principle of binary opposition, in terms of purity and pollution. There is an overarching impression that all castes, whether be high or low conformed to the single, as well as somewhat fixed textual hierarchy. But, there is considerable difference between what the textual view propounds and what actually takes place at the ground level. In fact, different castes groups adopt different strategies to interrogate the hierarchies and to establish their own notions of hierarchy and social standing. This explains the persistent attempts on part of 'lower castes', to repudiate the bonds of subordination that had been imposed on them, on the basis of the Hindu theory of *karma* and birth. [8] This is equally true in Indian army since the colonial period. Three Presidency armies followed different strategies to incorporate this concept. Among these, Madras army exceptionally maintained this policy to make army more viable and ladder for social mobilization for those particular caste groups who are in need.

After the Streynsham Master, the Yale's administration was one of military activity. Among the governor's earliest acts were placing of the native contingent on an improved footing. Records points out that there being 280 peons in pay for the watch and guard of the suburbs who in this disorder they were scattered about be of little credit or force to the place. He also passed order to call the Portuguese army and trained the peons and moor inhabitants according to their ability. [9] He also made rule and passed order for the military government. [10] The period of Harrison's rule was marked by considerable military activity. He appointed many peons for watching the black town. Records mentioned that apart from the official deployment, the occupational castes were employed for securities for the town. [11] The military establishment consisted in 1732 of a gunner, his first and second mates, 68 Europeans, 2 tindals and 27 lascars. [12] General Return of the troops in the Honourable East India Company's garrison of Fort St. George showed that the remarkable presence of the occupational caste groups and their power were utilised to capture and extend the British dominion in Indian subcontinent and overseas as well. [13] H. D. Love has referred about the various occupational caste groups, who were employed in the Madras army in 1759. [14]

Historians are divided on the issue of social implication of army on society. Some say that it has given a negative impact on the society because of their racial discrimination.

Some say that it was boon for the lower caste. The historian like Stephen P. Cohen is of the view that army was much beneficial for the Indian untouchables. [15] Meaning thereby caste and race, assumed colonial ethnographers, were the principal ingredients of Indian history, and had divided Indian society hopelessly. The new scholarly term for 'race' seems to be 'ethnicity' which was introduced into academic discourse by American political sociologists. For Stephen P. Rosen, Indian society is characterised by caste divisiveness. Hence, the colonial army remained an agglomeration of various castes. [16] In somewhat similar tune, DeWitt C. Ellinwood and Cynthia H. Enloe, following the historical sociological approach, argue that Indian society is ethnically fragmented. Ethnic identity is the product of primordial attachments resulting from religion, culture, language, and political influence. So, the sepoy army, assert Ellinwood and Enloe, Reflected ethnic imbalances. [17] For Rosen, divisions in the host society were automatically reflected in the army. Unlike Rosen, Enloe gives space to the programmes of the power elites. For Enloe, the degree and nature of ethnic imbalances in the armies were to an extent shaped by the politicians in power. [18] David Omissi notes that the ethnic make-up of the colonial army was not only shaped by the policies of the politicians and the generals, but also by the attitudes of the ethnic communities of the subcontinent. Only those ethnic groups who calculated that they would gain from military services joined the army.

So the ethnic composition of the colonial forces was the result of a fusion of the policies of colonial strategies and the dispositions of the various ethnic groups. [19] In a plural social set-up (in an ethnically heterogeneous society), ethnicity has enormous social and political importance. In a way, ethnic policies are the products of political manipulations as well as 'natural' divisions within a society.[20] The presence of ethnic communities with distinct characteristics, and jostling for power within the military bureaucracy, reminds one of Namierite factions. This is because the advocates of ethnicity accept that ethnic identities are closely related to interest-group orientations.[21] For the political sociologists, ethnic policies pave the way for political socialization and subsequent mobilization.[22] The persistence of ethnic groups which are leftovers from the pre-colonial state system even in modern armies somehow challenges the modernization theory.[23]

A group of scholars within the War and Society framework view that the army as a modernizing agency which transformed the peasant recruits into progressive individuals. This group could be categorized as the modernization school. The modernization theory became popular with American political scientists during the 1960s.[24] Ellinwood and S.D. Pradhan widen the analytical frame of the Modernization theory, and claim that modernization for global warfare not only modified the social relationship and mentality of the soldiers, but also transformed colonial society as a whole.[25] The modernization theorists occasionally view the army as modernizing the state apparatus and culture of the marginal groups such as women, lower castes, and untouchables.[26] In Cohen's view, during the emergencies the colonial army functioned as an instrument of modernization for the low castes.[27]

Oppressive caste discrimination and the practice of untouchability:

In India, the untouchable communities constitute a major segment of the Hindu society. For a fairly long period of time, they had been considered as 'untouchables' and had been physically segregated from the mainstream of all social transactions-the only exception being the use of their labour for production. Such a system of dominance had been sanctioned by the Hindu *Dharma Sashtras* and had been pursued with the same degree of rigour in all the regions of the country. Oppressive caste discrimination and the practice of untouchablility had characterised Indian society over the ages. [28] The all-India scenario remains to be the same while in case of the issue of untouchability since the colonial period. So, it was initiated to get more and more employment in the colonial army among the marginal communities or the untouchables in the eighteenth century onwards. In north India, they are termed as Dalit. [29] The southern part of the subcontinent, especially in Madras, they are termed as Paraiyans. In the past several decades, scholars dealing with 'untouchable' communities have focussed on a wide range of issues. In the first place, there had been an inclination to understand the reasons behind the origin and development of agrestic servitude in different parts of India. Scholars involved with such studies had been more interested in unearthing the specificities of the historical context in which the system of bondage emerged and developed some of its distinctive features. Secondly, in some other studies, scholars placed a greater deal of emphasis in understanding the factors responsible for the formation of distinct lower caste consciousness, role of caste associations and the political strategies pursued by the 'lower caste' politicians in the wake of the institutional concessions offered by the colonial government, as well as, by the governments of the native states. During the colonial period onwards, the usage of the terms such as national community and national culture reflects the story of upper caste domination over different subaltern social groups within the nation. There have been some studies, which have tried to understand the contemporary social behaviour of the 'untouchable' communities. Scholars engaged in such studies have tried to explore the attitudes of the 'untouchable' communities towards the prevailing caste system.[30]

The narrative on agrarian bondage should essentially begin with the opinions expressed by Dharma Kumar in this regard. In her research, Mrs. Kumar had argued that the bonded labourers belonged to the lowest rung of the Hindu castes and were in many places just 'untouchables' having no recognised position in the caste system. She argued that caste not only confirmed the social and economic disadvantages of the agricultural labourer, but also gave him some rights, some of which were economic, while others were of a social and ritual nature.[31]

Social composition of the Madras army :

Despite an initial bias for recruiting men of high caste origin, the Madras army was composed mainly of Muslims, middle-caste Hindus, some low castes, and only a few high castes.[32] For instance, the 1st Infantry regiment, which could be taken as a microcosm of the Madras infantry in 1824, was composed of 45 percent Muslims, 25.6 percent Telingas and 14.5 percent Tamils, 5 percent low castes, and 9.2 percent high castes. While most of the high castes came from Hindustan, (Purab, that is, west Bihar, Eastern and Southern Awadh/Oudh, especially the Bhojpur region) the other communities came from Karnataka, Mysore, and Andhra Pradesh.[33] Douglas M. Peers ascribes this preference for middle and low-caste soldiers to the fact that military service was not the preserve of high caste communities in south India.[34] Then too, the Bengal army monopolized the better sorts of higher-caste recruits from north India, leaving second grade recruits for the Madras army.[35] However, we feel that the induction of low and middle-caste recruits on part of the Madras army not entirely due to the non availability of high caste soldiers. There were deeper ideological underpinnings involved. Probably this decision to recruit low caste recruits was intended to isolate the Indian contingent of the Madras army from local attachments and sensibilities. General Harris, one of the main proponents of low caste recruitment, noted that the lack of religious prejudices and local attachments, among others, as qualities of the low-caste recruits.[36]

However, many caste and communities were taken as sepoy in the Madras army. Ardythe Basham, a Canadian scholar has taken up the study of untouchables in the western army, pointing out that Madras army consisted of low caste Hindus, untouchables, Christians converts. Besides, there were Pathan, Rajputs, Muslims from Northern states, Brahmins from Oudh, Sihar and Rohilkhand.[37] According to Stephen P. Cohen, Madras army exclusively recruited the Untouchables caste soldiers so that it could follow the orders of the colonial government. In this respect, Madras army employed the paraiyans caste and within the short time they earned the important place in the army.[38] the Paraiyans used to incorporate themselves in the colonial army so that they could explore their physical ability as well as earned the maximum respect in the society. However, the Streynsham Master, the earliest governor of Madras, was a person who made several discipline and rules for the Madras army. He gave order to train the all rank officers and paid heavy discipline.[39] This gave a new opportunity and new power to the occupational caste groups like the Paraiyans in the Madras army.

The Paraiyans and the Madras Army in the nineteenth century:

The advent of British rule in south India provided the Paraiyans and the kindred classes with opportunities of employment in the army. The Paraiyans since the 1760s and 1770s had constituted the bulk of the foot soldiers in the Company's army.[40] In the following decades too, they continued to find employments in the military department. The military depots, functioning in Madras and Trichinopoly, often served as recruitment centres for the Paraiyans. The British army bosses praised the Paraiyans recruits highly for their submissive nature and dutiful conduct. The high ranking British officials expressed the opinion that opportunities to serve in the army had inculcated among the 'untouchables' a certain degree of self respect and independence. The Paraiyans also sometimes expressed their satisfaction for being offered employments in the British army, as it provided them with the privilege to experience the civil equality enjoyed by the other British as well as Native subjects.[41] The extremely docile and loyalist attitude of the Paraiyan soldiers towards their British seniors, more than often accounted for their promotions.

Significantly, by the early decades of the nineteenth century, recruitment in the British army brought about important changes in the self-perceptions of the ordinary Paraiyan soldiers. In fact, they might have realised that sincerity and devotion could guarantee an improvement in their economic status.[42] Thus many of them adopted an overtly loyalist line to gain promotions to the ranks of non-commissioned officers in the regiments. The performance of the military rituals and drills also invoked in them a sense of belonging to a martial race. The award of the King Emperor's uniform instilled in them an idea that caste discriminations could no longer keep them tied to bonds of servitude and exploitation.[43]

The Paraiyans were exclusively recruited for one of the regiments of the Indian army, more popularly known as the '*Queens Own Sappers and Miners*' till about the middle of the nineteenth century.[44] But, after the Great Rebellion of 1857, there was a shift in the British Government's military recruitment policy. At this time, the British military superiors felt that the recruitment policy needed to be based on the 'martial race' theory. It was argued that while recruiting soldiers from the various native communities, the military, social; and environmental perspectives needed to be taken into account. In other words, different aspects of the martial race theory were employed to recruit native soldiers for the Indian army. Jats, Gurkhas, Sikhs and Pathans were recruited in large numbers in the Bengal, Bombay and the Madras armies.[45] These communities were believed to be more trustworthy, tough and hard working than the native soldiers belonging to the other communities.

Changing attitudes of the colonial British military officials:

In view of these changes in the recruitment policy, there was an appreciable decline in the number of Paraiyan soldiers. In the 1870s, the British military officials favoured the recruitment of upper caste Hindus, Christians and Mussalmans.[46] However, the Paraiyans did have an impressive presence in some of the branches of the military department. In fact, as late as the 1890s, they enjoyed virtual monopoly as sappers and miners. The upper caste apathy for menial occupations was partly responsible for the impressive presence of the Paraiyans in certain branches of the army.[47]

In the early 1890s, the disbandment of the old Presidency armies as well as the increasing recruitment of North Indian martial races denied the Paraiyans further opportunities of future recruitments in the army. The Paraiyan community leaders were greatly disturbed by these developments. Most of them believed that military employment could act as a channel for their economic and social mobility. The memoranda and petitions sent to the British government by the Paraiyan community leaders like Pandit C. Iyothee Thoss clearly brought out the importance that the community members attached to military service. In a memorandum sent to a government official in 1894, Iyothee Thoss emphasised that government intervention was needed to improve the conditions of the 'untouchables', since they had displayed great valour and sacrifice as soldiers. [48] But, Iyothee Thoss was not the lone person to raise the issue of preferential treatment in lieu of the long standing military service offered by the members of the socially despised communities. The small coterie comprising of Paraiyan ex-servicemen and army pensioners, also raised this issue to establish themselves as the real representatives of the community. Thus, it could be argued that a long standing involvement with the British Indian army served as one of the key elements behind the growth of Paraiyan community consciousness in the Madras Presidency. [49]

Conclusion:

The Madras army has given much benefit to the occupational caste groups. It becomes a source of social occupational mobility. It gave a new identity and new power. Their caste becomes empowered. They got education and trained from the English which produced some marshal castes like Paraiyan become much empowered. Army were however, quasi-mercenary forces commanded by a foreign group of officers, who had established their sway by force. Consequently, the question of national loyalty did not arise. Loyalty in a quasi-mercenary army could only mean loyalty towards commanding officers who, by exhibiting dynamic leadership, sought to win allegiance. The British initially tried to ensure personal loyalty by encouraging a very personal and heroic style of leadership. With the transition to an impersonal style of leadership, both the armies suffered to an extent, especially in times of crisis caused by contractual violations, flouting of long standing traditions, and so on. Consequently, in mutiny like situations where the privileges, both the armies were extremely vulnerable and mutinies were often marked by a total collapse of the command mechanism. While the Indian officers sided with European officers in maintaining day-to-day discipline, they could not be relied upon in times of acute or mass discontent. Then the European officers were left helpless spectators.

Notes :

1. I am thankful to my guide Prof. Raj Sekhar Basu who has inspired me to write an article on this topic. After the release of his book, '*Nandanar's Children: Paraiyans' Tryst with Destiny, 1850-1956,* New Delhi, it gives me an opportunity to write this article. I am indebted to him.
2. The Paraiyans who were scattered over a large part of Tamil Nadu comprised the largest 'untouchable' agricultural labouring community in the early nineteenth century Madras Presidency. It was widely believed that they derived their name from the Tamil word *Parai* , which usually denoted a musical instrument resembling a drum or a tom-tom. In their writings, the English officials frequently referred to them as community of drummers, who performed during the funerals and village festivals. The term Paraiyan, in terms of a caste, or more correctly an occupational nomenclature is believed to have first appeared in a poem of Mangudi Kilar of the second century AD.
3. Kaushik Roy, ed., *War and Society in Colonial India,* Oxford University Press, New Delhi, 2006, p. 1.
4. Ibid., p. 4.

5. Ardythe Basham, *Untouchable Soldiers: The Mahars and the Mazhbis*, edited by Bhagwan Das, Gautam Book Centre, New Delhi, 2008, p. 19.
6. Ibid., p. 19.
7. Dipankar Gupta, *Interrogating Caste: Understanding Hierarchy and Difference in Indian Society*, New Delhi, 2000, p. 1.
8. Dipankar Gupta, Towards Affirmative Action, in Ira Pandey, ed., India 60, India International Centre, Quarterly, Winter 2006 and Spring 2007, New Delhi, 2007, p. 153.
9. Records of Fort St. George, Diary and Consultation Book, 1778, Tamilnadu Archinex (henceforth TNA), p. 107.
10. Ibid., pp. 108-9.
11. Records of Fort St. George, Diary and Consultation Book, 1714, TNA, p. 133.
12. Records of Fort St. George, Diary and Consultation Book, 1732, Tamilnadu Archives. This describes the various duties of the occupational castes groups. These are as follows- 8 persons were disposed at the inner Fort gunroom, 21 were at the Saluting battery, 12 were at the new Power house, 9 were at the Garden point, 6 were at the Queen point, 2 were at the bridge point, 2 at the spur point, 2 were at the Clark point, 2 were at the Clark gate, 2 were at the Madepollam point and 2 were Fleet point and the total was 68 persons who were in charge of the all round security of the garrison.
13. The return shows that the presence of the occupational caste groups. The numbers are as follows- Troops of Horse were 35, Royal Artillery were 132, the Honourable Company's Artillery were 64, His Majesty's 64th Regiment were 195, the Detachment of Marines were 100, the Honourable Company's 1st Battalion were 625, the Honourable Company's 2nd Battalion were 375, Supernumeraries were 32 in numbers and the total was 1758 where as the officers were in these said Battalions were only 94. For more detail, see Records of Fort St. George, Diary and Consultation Book, 1749-50, p. 32 and see the list of servants in the page no. 202 in the same record, TNA.
14. H. D. Love, *Vestiges of Old Madras 1640-1800*, Vol. 2, Asian Educational Services, New Delhi, 1996, p. 140.
15. Stephen Peter Cohen, Indian Army, its contribution to the Development of a Nation, Oxford University Press, Bombay, 1990, p. 13.

16. Stephen Peter Rosen, *Societies and Military Power: India and Its Armies*, Oxford University Press, New Delhi, 2006, p. 5.
17. DeWitt C. Ellinwood and Cynthia H. Enloe, 'Preface', in Ellinwood and Enloe, ed., *Ethnicity and the Military in Asia*, Transaction Books, USA, 1981, p. 2.
18. Cynthia H. Enloe, 'Ethnicity in the Evolution of Asia's Armed Bureaucracies', in Enloe and Ellinwood, ed., ibid, pp. 2-14.
19. David Omissi, 'Martial Races: Ethnicity and Security in Colonial India, 1858-1939', War and Society, Vol. 9, No. 1, 1991, pp. 2-27.
20. Zakaria Haji Ahmed, 'The Bayonet and the Truncheon: Army and Police Relations in Malaysia', in Ellinwood and Enloe, eds., ibid., pp. 209-10.
21. Lancy Bruce Fields, 'Ethnicity in Tso Tsung-T'ang's Armies: The Campaigns in North West China, 1867-80', in Ellinwood and Enloe, eds., Ibid., pp. 55-74.
22. William R. Heaton, 'The Chinese People's Liberation Army and Minority Nationalities', in Ellinwood and Enloe, ed., Ibid., p. 177.
23. Constance M. Wilson, 'Burmese-Karen Warfare, 1840-50: A Thai View', in Ellinwood and Enloe, ed., Ibid., 19.
24. For a discussion on these aspects see Jonathan R. Adelman, 'The formative Influence of the Civil Wars: Societal Roles of the Soviet and Chinese Armies', Armed Forces and Society, Vol. 5, No. 1, 1978, pp. 93-116.
25. DeWitt C. Ellinwood and S.D.Pradhan, 'Introduction', in Ellinwood and Pradhan, *India and World War I*, Monohar, New Delhi, 1978, pp. 2-18.
26. Carol Hills and Daniel C. Silverman, 'Nationalism and Feminism in Late Colonial India: The Rani of Jhansi Regiment, 1943-45', Modern Asian Studies (henceforth MAS), Vol. 27, No. 4, 1993, pp. 742-60.
27. Stephen P. Cohen, 'The Untouchable Soldier: Caste, Politics, and the Indian Army', Journal of Asian Studies (henceforth JAS), Vol. 28, No. 3, 1969, pp. 453-68.
28. Raj Sekhar Basu, *Nandanar's Children: The Paraiyans' Tryst with Destiny, Tamil Nadu 1850-1956*, Sage Publications, New Delhi, 2011, p. xxiv.
29. The term Dalit in popular parlance represents a wide range of meanings, all of which convey a tale of exploitation and discrimination. See, Sagarika Ghosh, The Dalit in India, Social Research, Vol. 70, No. 1, 2003, p. 86.

30. Raj Sekhar Basu, Reinterpreting Dalit Movements in India, Indian Historical Review, Vol. 3, No. 1, 2006, pp. 72-73.
31. Dharma Kumar, Land and Caste in South India: Agricultural Labour in the Madras Presidency During the Nineteenth Century, (Reprinted with an Introduction), Cambridge University Press, New Delhi, 1992, p. 34.
32. Henry Russell, *Select Committee for Indian Affairs, 1831-32,* Minutes of evidence, Vol. 13, p. 567, TNA.
33. Henry Dodwell, *Sepoy Recruitment in the Old Madras Army,* Superintendent of Government Press, Calcutta, 1922, Appendix, p. i.
34. Douglas M. Peers, 'Habitual Nobility of Being: British officers and the Social construction of the Bengal Army in the Early Nineteenth Century', MAS, Vol. 25, Issue 3, 1991,pp. 551-2.
35. Kaushik Roy, 'Recruitment Doctrines of the Colonial Indian Army, 1859-1913', Indian Economic and Social History Review', Vol. 34, No. 3, 1997, pp. 345-6. Here he argues that a lobby existed in the pre-1857 madras army, which pursued an open door policy of recruitment. General Harris was one of the prominent proponents of the open door policy won against the selective enlistment school due to certain factors. One of the reasons was that since Awadh and Bengal Army had the first choice of Brahmin and Rajput recruits in these areas. As a result, the Bengal Army acquired the better variety of Purbiya recruits, leaving the second grade high caste personnel for the Madras army. So the Madras army decided to go for the non-high caste men.
36. W. J. Wilson, History of the Madras Army, Vol. 3, Govt. Press, Madras, 1883, p. 151.
37. Ardythe Basham, *op. cit.*, p. 9.
38. Stephen Peter Cohen, op. cit., pp. 453-68.
39. Records of Fort St. George, Diary and Consultation Book, 1678, p. 122. See also, Diary and Consultation Book, TNA, 1679.
40. T. P. Kamalnathan, *K. Veeramani is refuted and the Historical facts about the Scheduled Caste Struggle for Emancipation in India,* Tiruppathur, 1985, p. 35.
41. Calcutta Review, Vol. 33, 1859, p. 143.
42. Raj Sekhar Basu, *op. cit.*, p. 160.
43. Ibid., see also V. Geetha and S.V. Rajadurai, *Towards a Non-Brahmin Millenium: from Iyothee Thoss to Periyar,* Popular Prakashan, Calcutta, 1998, p. 72.
44. T. P. Kamalnathan, *op. cit.*, p. 34.

45. For more details, see Stephen P. Cohen, "The Untouchable Soldier: Caste, Politics of the Indian Army", JAS, Vol.XXVIII (28), No. 3, May 1969, p. 456.
46. Madras Army Records, Box. 10, Series-II, January 1872, TNA cited in Raj Sekhar Basu, *op. cit.*
47. T. P. Kamalanathan, K. Veeramani is refuted, p. 34.
48. Pandit C. Iyothee Thoss, Open letter to the Hon'ble Srinivasa Raghavaiyangar (Inspector General of Registration), Madras, 1894, Theosophical Society Library, Adyar, p. 4 cited in Raj Sekhar Basu, *op. cit.*
49. Raj Sekhar Basu, *op. cit.*

Black Peepers who charged : Remembering the British-Indian Military Personnel of the Two World Wars

Pinaki Roy

> Theirs not to make reply,
> Theirs not to reason why,
> Theirs but to do and die
>
> Alfred, Lord Tennyson
> 'The Charge of the Light Brigade' (1854)

In *Our Armed Forces*, Gautam Sharma recounts how within a few years after the 1857 Mutiny, there was a renewal of demands for greater participation of Indians in the British Indian Army. [1] While this may apparently suggest the outstanding valour usually exhibited by the Indian recruits of the colonisers' military force, the English keenness on drawing more subaltern in their forces also signify the Europeans' habit of passing the main burnt of belligerence to their forces comprised of the 'expendable' colonised people. Ingloriously, and sometimes forcibly drafted, the Indian fighters found themselves in a situation not unlike that of the English cavalry soldiers who were ordered to charge at the conveniently-placed and well-armed Russians during the 25 October 1854 *Battle of Balaclava*, commemorated by Tennyson in his famous "The Charge of the Light Brigade". These subaltern recruits could not question their English superiors, nor resent their motives they could only fight courageously and perish miserably.

Indeed, during both the World Wars, the Indian volunteers formed the largest past of the imperial British Army. [2] As Nayar and Paul point out, having had contributed 1.3 million troops during the 1914-18 war and 2.5 million personnel during the 1939-45 global belligerence to the Allies, the Indians created the largest volunteer army in history. [3] During the First World War, the Indian section of the Allied forces won 13000 medals, including 12 *Victoria Crosses*; in the Second, there were at least 4000 decorations, including 38 *Victoria* and *George Crosses*. In the period of 1914-18, a total of 74,187 Indian soldiers were killed, including 43,000 in direct action; between 1939 and 1945, approximately 36,000 Indian members of the Allied troops sacrificed their lives. Also, during the two World Wars, the approximate total number of grievously-mutilated Allied Indian military personnel was 0.1 million. Such statistical figures suggest that in spite of being members of the subaltern, Indian military personnel never lacked courage or loyalty while they defended their English rulers and their allies albeit almost always confused about their own advantages from such confrontations. However, such soldiers experienced apartheid while in service. They were often poorly equipped, [4] and, as Bridger writes, were strictly supervised by White British officials. [5] It has also been often alleged that the Indian fighters of the Allied troops were segregated on the basis of tribes, castes and religions during both the World Wars.

During the First World War, as a part of the *British Expeditionary Force A*, the Indian soldiers were meted out demeaning and often disastrous treatments by their Allied recruiters. Upon their arrival in Marseilles on 30 September 1914, only six weeks after the declaration of war, the members of the *Lahore Division* were forced to take part in the inconclusive *Battle of La Bassée* (10 October 2 November 1914) in northern France. It was followed on 10-13 March 1915 by the *Battle of Neuve Chapelle* in the French Artois region, commemorated by John Fowles in *The Magus* (1966). During both these confrontations, the Indian casualties were considerably high. In the *Battle of Neuve Chapelle* alone, approximately four thousand and two hundred Indians were killed by the German military initially under the command of General Erich von Falkenhayn. In spite of the 1902-09 modernising efforts by Horatio Kitchener, 1st Earl Kitchener of Khartoum, the British Indian militarymen remained unacquainted with *Charger Loading* and *Short Magazine Lee Enfield Mk I, II, III* and *V* rifles, which were issued to them only after their landing in Europe. They were denied artillery support. Underdressed against the severe French chill, they were often arbitrarily drafted in from unrelated and separately-positioned regiments. [6] Their White commanding officers could not speak their languages and often misdirected military assaults. [7] It may be recalled that discriminating tactics followed by the White *East India Company* officers had resulted in the *First Indian War of Independence* derided by the English colonisers as the 'Sepoy Mutiny' (10 May 1857 20 June 1858). As Fremont-Barnes in *Essential Histories: The Indian Mutiny1857-58* and the "1857 Indian Rebellion" section of the *Victorian Web* website (available on *http://www.victorianweb.org/history/empire/1857/index.html*) sum up, such disastrous policies included denial of proper pay, pensions and promotions to the army personnel, particularly those belonging to the Bengal brigade, sending them overseas to fight against their religious objections, and issuance of *Pattern 1853 Enfield* rifles that used cartridges most probably covered with cow and pig fat-smeared papers. [8]

As Michael Duffy records in "Primary Documents: German Report of the *Battle of Neuve Chapelle*" (2009), Margarete Munsterberg, a German journalist reporting the confrontation in the Berlin periodical *Kriegs-Rundschau*, thus reports the Indians:

> "[T]he fight over Neuve Chapelle continued. Here Indian troops rushed ahead and seemingly unarmed. In the preceding days numerous Indians had deserted to our lines, hence our troops believed that in this case they were again dealing with deserters and so did not shoot. This sin of omission was thoroughly avenged; for close before our positions the Indians began to throw hand grenades and attacked the garrison of our trenches with knives".

Moreover, as Gardner points out, the Indian soldiers serving at Marseilles were kept virtually under confinement for their recruiters' fear about their establishing sexual relationships with White women, and they "reacted negatively to their lack of freedom in France".[9] An Indian army personnel, in a 2 December 1915 letter addressed to his family members, writes,

> "Alas we are not free to go about at will. In fact we Indians are treated like prisoners. On all sides there is barbed wire and a sentry stands at each door. Leave London out of the question; we cannot even get to see New Milton properly. If I had known that such a state of affairs would exist, I would never have come. If you ask me the truth, I can say that I have never experienced such hardship in all my life. True, we are well fed, and are given plenty of clothing but the essential thing freedom is denied. Convicts in India are sent to Andaman Islands; but we have found our convict station here in England".[10]

During the Second World War, the problem of poor supply of equipments to the Indian members of the Allied forces persisted, regardless of the fact that, as Shirer writes in *The Rise and Fall of the Third Reich*, the British colonisers considered protecting their Indian colony and seeking protection from the sub-continental colonised recruits as priorities.[11] As Sumner writes,

> "The Indian Army entered the Second World War woefully underprepared. The amount of money devoted to the army by the Government of India had slowly declined after the end of the First World War. Despite a subsidy from Great Britain it was left desperately short of modern weapons and equipment and it did not possess sufficient men able to handle mechanical transport [...]. Only with the recruitment of a large number of Madrassis and Mahrattas did this situation begin to change. Neither the Japanese depredations in China nor the outbreak of war in Europe disturbed the sanguine outlook of the British and Indian governments indeed, the advice coming from London was that it was unlikely that Indian troops would be required at all. The extent of the governments' miscalculation was apparent when, at the outbreak of war, the Indian Army quickly found itself in the field again, immediately contributing two divisions [...] to the operations in the Western Desert and Abyssinia [...]".[12]

The Indians, who had all along been informed that given their underprepared state they were unlikely to be pitted against the highly-mechanised and organised German, Japanese and Italian forces, were quietly sent to the North and East African campaigns. To recall the difficulties that soon followed, the *31st Armoured Division*, to exemplify, severely lacked in tanks,[13] withered mules and light vehicles were used to transport weaponry and equipments to the *17th and 39th Indian Infantry* divisions,[14] while the artillery units lacked in long-range howitzers. Even during the Second World War, the Indian militarymen had to serve in some of the fiercer Allied military operations. The 4th *Indian Infantry Division* participated in *Operations Compass* (against the Italians at Sidi Barrani-El Agheila, 8 December 1940-9 February 1941), *Battleaxe* (against the combined German-Italian forces at Cyrenaica, 15-17 June 1941), *Crusader* (against the combined German-Italian forces in Egypt and Libya, 18 November-30 December 1941), and *Pugilist* (against the combined German-Italian armymen in southern Tunisia, 16-27 March 1943) in North Africa. In August 1940, the British Indian armymen fought against the Italian Army during the *Fascist Conquest of the British Somaliland* (August 1940). They participated in the *Anglo-Iraqi War* (against the Arabian-Iraqi soldiers of Rashid Ali, 2-31 May 1941), the *Syria-Lebanon Campaigns* (against the forces of Vichy France and Germany, 8 June-14 July 1941), the *Battle of Hong Kong* (against the Japanese, 8-25 December 1941), *Battles of Malaya* and *Singapore* (against the Japanese, 8 December 1941-31 January 1942 and 31 January-15 February 1942, respectively), and particularly, in the *Burma Campaign* (against the Imperial Japanese armymen, January 1942 July 1945). However, their conditions never improved, and they continued to be treated as subservient to the White members of the Allied forces.

This essay tries to offer a few glimpses into the psychological turmoil the Indian militarymen experienced during the segregating and humiliating days of the World Wars. The perplexing conceptions of 'Hindustan', 'British India' and 'India' could not have been more suitably explored than a reference to the bewilderment of these south-Asians. They showed steady allegiance, as David Omissi writes in "*World Wars in Depth*: India and the Western Front", to the English king, George V, but were often infuriated by the demeaning attitude of their White superiors.[15] Dudley *et al.*, referring to William La Follette's address to the *U.S. House of Representatives*, view this as a situation that must arise when a subaltern tries to collaborate with an imperial force which has a history of most brutal suppression of the same colonised members. [16] Naturally, the courage and self-sacrifices of the Indian soldiers have scarcely been remembered by the White Western historians and litterateurs. In English literature, among other very limited portrayals, they are briefly commemorated in the undated and uninspiring sixteen-liner "Chindit" by K.N. Bateley the poem's title referring to the Orientalist practice of the White officers of *77th Indian Infantry Brigade* (commanded by Brigadier Orde Wingate) who called the long-range-penetrating Indian soldiers fighting in Burma after mythical beasts guarding Burmese temples. Ironically, during *Operation Longcloth* in February 1943, these 'Chindits' displayed courage far more than their White superiors. [17] A glimpse at the Mary Hammond-and-Shafquat Towheed-edited *Publishing in the First World War: Essays in Book History* (Basingstoke-Hampshire : Palgrave-Macmillan, 2007) would reveal the absence of quality European writings eulogising the Indian participation in the World Wars.

For the British Indian soldiers, perplexity and indignation increased during two specific sections of the First and Second World Wars. During the First World War, by end-October 1914, Turkey had effectively joined in on side of the Central Powers. Sensing that the British Empire was then at war with an Islamic power, the Muslim soldiers of the British Indian Army began to demonstrate their resentment. For long, religious artefacts and observations of ceremonies had been most important to the Hindu, Muslim and Sikh soldiers of the British Indian Army.

Earlier, religious and caste issues were responsible for fuelling the 1857 rebellion. [18] The post-1914 anti-Islamic operations were perceived by the Muslim soldiers, who had a sizeable presence, as desecrating their religiosity. In letters written from the front, many of which have been collected in Omissi's *Indian Voices of the Great War*, some soldiers are found cleverly registering their protest against their White Christian commanders. [19] The Allied Islamites, especially those deployed on the Western front, began defecting or deserting, while there were at least three mutinies among soldiers fighting on other fronts. Ferro offers the picture of a rather interesting situation at this juncture :

> "[The] war encouraged anti-colonial movements where they had begun, and started them where they had not. In India [...] there was, despite noisy demonstrations of loyalty, little sympathy for the British. German victories caused rejoicing among both [Indian] moderates and extremists [...]". [20]

Gordon Corrigan, while focussing on the misery of Indian soldiers placed in trenches throughout his book *Sepoys in the Trenches: The Indian Corps on the Western Front* (The Mill Brimscombe Port Stroud-Gloucestershire: Spellmount Publishers, 1999) writes about how the resenting soldiers were intermittently causing self-inflicted wounds and showing reluctance to fight. As Leonard Shurtleff exasperatedly writes in a 1998 e-article "France at War: Tirailleurs Senegalais",

> "Britain's own experience with Indian Army troops on the Western Front was not a good one. Indian troops were employed there only briefly and had a high rate of self-inflicted wounds. Some Indian units mutinied when they learned they were to be sent outside the subcontinent. Unlike French colonial troops, they had not been told to expect worldwide service. They were not trained or equipped for

> service outside India and it was never intended that they would serve in Europe. And in the Second War they did not while French African troops again served on the Western Front".

To the Indians, many of who had been forcibly recruited and some of whom later led the anti-British movements in Punjab in 1919, 'Hindustan' had long been changed into 'British India'; but they were fighting to free India not from the colonisers but from enemies of those imperialists who have had so long been so brutal against them. During both the World Wars the Indians were objectionably underpaid compared to their White comrades David Omissi opines in his *B.B.C.* article that an Indian soldier's average monthly pay during the First World War was eleven rupees. However, they fought for the modest payment, in hopes of promotion, and for bringing glory to their clans and castes. [21] Such motivations, nevertheless, were flagging by end-1914.

During the Second World War, the Indian soldiers discovered themselves as being more frequently pitted against the comparatively fiercer native freedom fighters than the White colonisers. They had to withstand North African heat of deserts, were almost always short of ration and equipments, and had to be constantly on guard watching for the Axis armymen while their White superiors rested and cavorted. Many of the British Indian soldiers, as Satyendranath Basu writes in *Alongwith the Azad Hind Fauj,* had been influenced by the contemporary Indian politicians and their ideologies, and a considerable section of these armymen was soon showing reluctance to fight.[22] More importantly, as Bipan Chandra *et al.* write in *Freedom Struggle,* the domination and coercion of the British colonisers had become intolerable by the beginning of the Second World War, and this may have given considerable pangs of conscience to the Indians fighting on side of the

Englishmen.[23] However, regardless of what Basu writes, many of the Indian politicians, particularly those belonging to the *Indian National Congress* party, added to the confusion of these soldiers. As Madhu Limaye and Rajendra Dube develop respectively in their (four-volume) *Mahatma Gandhi and Jawaharlal Nehru: A Historical Partnership 1916-1948* (New Delhi: B.R. Publishing Corporation, 1989. Rpt. 1990) and *Jawaharlal Nehru: A Study in Ideology and Social Change* (New Delhi: Mittal Publications, 1988), the *I.N.C.*, led by Mohondas Karamchand Gandhi, Vallabhbhai Patel and Abul Kalam Azad, took a strong stance against Fascism and Nazism as sources of violence. Several Indian leaders supported the British cause, though Jawaharlal Nehru pointed out the inherent contradiction in the British argument of going to war with Nazi Germany for the sake of freedom, since India was denied that same freedom. In August 1942, the *Indian National Army* was formed in Singapore, and in 1943, Subhas Chandra Bose took over as its leader. When locked in armed conflicts with the approximately forty-three-thousand-strong cadres of the *Indian National Army*, a considerable section of the British Indian Army either surrendered out of will, defected, or was overpowered by the valorous freedom-fighting soldiers. In the partly-fictionalised *Netaji I have met: Confessions by Witnesses*, Narayan Sanyal traces in details the psychological turmoil faced by the British Indian soldiers when confronted over the issue of fighting against the *Indian National Army* volunteers.[24] Mukhopadhyay opines that Subhas Chandra Bose was the only Indian leader who had earned the respect of both the Indian and British Indian Armies,[25] and his pro-independence leadership had, identifiably, accentuated the psychological pressures of the British Indian soldiers.

The Indian recruits of the imperial army were also infuriated by the military censorship particularly during the First World War. Having had been wounded at the front, numerous Indians were transferred to England to recuperate while at the same hospitals were posted their own countrymen serving as medical personnel, cooks, storekeepers, general clerks, and so on. English officers, scared about the distribution of revolutionary pamphlets among these soldiers, set up a censors' office at Boulogne under E.B.Howell, and the recovering Indians soon intercepted this secret vigilance. Already suffering from 'divided hearts' whether to fight for or against the British imperialists they understandably began to nurse resentments. This issue, addressed by Omissi in his *Indian Voices of the Great War: Solders' Letters, 1914-1918* and *The Sepoy and the Raj: The Indian Army, 1860-1940* (Basingstoke-Hampshire: Palgrave-Macmillan, 1994), led to the Indian recruits' designing of 'literary camouflages' to befool their imperial supervisors. Though numerous Indian soldiers of the imperial army, mostly recruited from rural areas, were illiterate and lacked capacities of writing letters, they devised indigenous coded signs and languages to let their family members and acquaintances know about their miserable conditions and resentments for example, substituting 'black pepper' to indicate the natives of the British Indian Army. Interestingly, such desperateness to write home forced many Indian armymen to take self-education and enrich themselves in rural and agrarian imageries to elude the White censors. Rozina Visram's "The First World War and the Indian Soldiers", published in *Indo-British Review: A Journal of History* (Volume XVI, Number 2, June 1989) offers a historical inquiry into the soldiers' letters and their wartime sentiments, including when convalescing at the *Indian Military Hospital* at Royal Pavilion, Brighton.

In his "The Imperial Reserve", Jeffrey Greenhut, while commemorating the courageousness of the Indian soldiers, also perceives them as people who are frequently in despair about returning home. [26] Perhaps, because of this sense of temporality of life, the Indian soldiers tried to discover Europe as much as possible in the days they were posted for training or to the front, or convalesced at army hospitals. They admired the intelligence, generosity, honesty and integrity of the Europeans, and tried to impose some of the European standards having had returned back to their own places. However, many Indian recruits also tried to avoid Europe because according to them, the continent might have been full of beautiful and large buildings and thriving industries, but it lacked spirituality and was 'frankly immodest'. Philippa Levine, in her "*Battle Colours*: Race, Sex, and Colonial Soldiery in the First World War", not only explores the apartheidic practices of the English soldiers towards their Indian comrades but also focuses on the Indian soldiers-European women relationships. [27] About this, Gardner writes in details:

> Of particular concern to British authorities was the prevention of sexual relations between Indian soldiers and European women. As Greenhut has observed, service in France provided the Indians with unprecedented contact with women of European descent. The prospect of intimacy between the two groups offended the racist sensibilities of the British. In addition, colonial authorities feared that sexual contact with European women would undermine the esteem in which Indian soldiers ostensibly held British women in India, thereby threatening the foundations of colonial rule". [28]

Simultaneously praising the White women for their education and courageousness and criticising them for their 'shamelessness', many of the Indian soldiers transported to Europe established sexual relationships with particularly British and French women during the First World War, and with British, German, and Japanese women during the Second World War. Even Muslim-Christian marriages occurred, as David Omissi reports in his *B.B.C.* article "World Wars in Depth: India and the Western Front". Following their marriages to the European women, the British Indian soldiers came to suffer from pangs of conscience because they thought that they had befooled not only their countrymen but also their family members.

This essay has been prepared to briefly explore the physical and mental conditions of the British Indian soldiers principally serving overseas during the two World Wars. It must be mentioned that the issues of 'India' and 'British India' especially intrigued the Indian recruits of the imperial army, many whom ended their service life without recognising the real enemy. They were almost always segregated from the White British army personnel and were stereotyped by their dresses and colours. [29] Das writes about how the White Anglo-French soldiers fighting during the First World War always remained in anxiety lest their interactions with the African and Indian armymen made them 'racially impur[e]'. [30] Marika Sherwood is of the opinion that without active help from the Indian recruits, the Britons would have never won their part of the Second World War. This happened regardless of the fact that the 'colour bar' remained though unobtrusively throughout the war and that the colonised individuals had to take less pay, build their own barracks, and unceremoniously dumped and repatriated sometimes before their 'contracts' ended. Thus, studying the positions of the British Indian Army personnel during the World Wars would help readers recall the dichotomies of 'India' and 'British India' in the pre-independence days. As it was later revealed when historians began to focus in details on the shabby treatment of the British-Indian fighters at the hands of their imperialist recruiters, 'British India' was never an integral part of 'India' it was rather an appropriation of the vast south Asian country into a colony which would cater to the Orientalist perspectives of the White Westerners and allow them to test their notions of the sub-humanity and primitive defencelessness of the Easterners.

Notes :

1. Sharma Gautam. *Our Armed Forces*. New Delhi: National Book Trust, 2000. Rpt. 2006. 18-9.
2. Sumner, Ian. *The Indian Army 1914-1947*. Oxford: Osprey Publishing, 2001. 3.
3. Nayar, Baldev Raj, and T.V.Paul. *India in the World Order: Searching for Major-Power Status*. Cambridge: Cambridge University Press, 2003. 66.
4. Sumner, Ian. *op. cit.*, 5.
5. Bridger, Geoff. *The Great War Handbook*. Barnsley: Pen and Sword, 2009. 63.
6. Sumner, Ian. *op. cit.*, 5.
7. Ibid.
8. Fremont-Barnes, Gregory. *Essential Histories: The Indian Mutiny 1857-58*. Oxford: Osprey Publishing, 2007. 25.
9. Gardner, Nikolas. *Trial by Fire: Command and the British Expeditionary Force in 1914*. Santa Barbara-C.A. and London: Praeger Publishers, 2003. 177.
10. "British Library's *Asians in Britain*: Indian Soldiers in World Wars". Accessed on 1 September 2010. *<http://www.bl.uk/reshelp/findhelpsubject/history/history/asiansinbritain/indiansoldiersinworldwars/indiansoldiers.html>*
11. Shirer, William L. *The Rise and Fall of the Third* Reich. New York : Fawcett Crest, 1962. Rpt. 1983. 1155, 1178.

12. Sumner, Ian, *op. cit.*, 23.
13. Perry, Frederick William, *The Commonwealth Armies: Manpower and Organisation in Two World Wars,* Manchester: Manchester University Press, 1988, 11-14.
14. Jeffreys, Alan, and Duncan Anderson. *The British Army in the Far East 1941-45.* Oxford: Osprey Publishing, 2005. 23.
15. Omissi, David. "*World Wars in Depth*: India and the Western Front". Accessed on 20 August 2010. <*http://www.bbc.co.uk/history/worldwars/wwone/india_wwone_01.shtml*>
16. *The First World War: Opposing Viewpoints.* Ed. Dudley, William. San Diego: Greenhaven Press, 1998. 98.
17. Slim, William. *Defeat into Victory.* Cutchogue-New York: Buccaneer Books, 1956. 38.
18. Mazumder, Rajit Kumar. *The Indian Army and the Making of the Punjab.* New Delhi: Permanent Black, 2003. 7-8.
19. Omissi, David. *Indian Voices of the Great War: Soldiers' Letters, 1914-1918.* Basingstoke-Hampshire: Palgrave-Macmillan, 1999. 10-300.
20. Ferro, Marc. *The Great War.* Trans. Stone, Nicole. London and New York: Routledge, 2002. 116.
21. "B.B.C. World Service People and Beliefs: *Like another Mahabharata*: Indian Soldiers in the Great War" (2000). Accessed on 1 September 2010 <*http://www.bbc.co.uk/worldservice/people/highlights/001115_india.shtml*>
22. Basu, Satyendranath. *Alongwith the Azad Hind Fauj.* Kolkata: Ananda Publishers Private Limited, 1996. 43.
23. Chandra, Bipan, Amales Tripathi, and Barun De. *Freedom Struggle.* New Delhi: National Book Trust, 1972. Rpt. 1993. 192-214.
24. Sanyal, Narayan. *Netaji I have met: Confessions by Witnesses.* Kolkata: Deys' Publishing House, 1970. Rpt. 2005. 194-5.
25. Mukhopadhyay, Bibekananda. *The History of the Second World War.* Vol. II. Kolkata: Nabapatra Prakashan, 1986. Rpt. 1998. 259-72.
26. Greenhut, Jeffrey. "The Imperial Reserve: The Indian Corps on the Western Front, 1914-15". *Journal of Imperial and Commonwealth History,* 12 (1), October 1983: 54-73.
27. Levine, Philippa. "*Battle Colours*: Race, Sex, and Colonial Soldiery in the First World War". *Journal of Women's History,* 9 (4), Winter 1998: 104-30.

28. Gardner, Nikolas. *Trial by Fire: Command and the British Expeditionary Force in 1914*. Santa Barbara-C.A. and London: Praeger Publishers, 2003. 177.
29. Sharma Gautam. *Our Armed Forces, op. cit.*, 36.
30. Das, Santanu. *Touch and Intimacy in First World War Literature*. Cambridge: Cambridge University Press, 2005. Rpt. 2008. 59.

RESISTANCE

Protest Movements and Local Sources : A Study of Two Popular Protest Movements in Maldah 1938-1947

Ashim Kumar Sarkar

One of the very recent trends of the Indian historiography is that a number of historians have concentrated on writing the '*history from below*' i.e. the history of the toiling people of our country, through the ages. The ruling elite, not only of our country, but throughout the world, denied the past activities of the common people and discarded them from the arena of history, with a view to impose state's systems of control over the past. Hence, an attempt to write people's history has emerged as an effective answer to the traditional approach of writing elite history. In this context, we would discuss some popular protest movements and tribal resistance that emerged from the local conditions.

Our principal sources for information in this respect are Gourdoot and *Maldaha Samachar*, two contemporary local newspapers and the unpublished '*memoir*' of the editor of *Gourdoot*, Lalbihari Majumdar. Newspapers, local or national, is definitely a very viable source of history writing. The Marxist scholars on social history of Great Britain and France argued that oral evidences as well as newspaper materials could effectively counter the official source material preserved in archives or in several government reports. A vivid study of the newspapers might have been a fruitful corroborating source which could act as checks and balances of ruling class documents. We would try to present a systematic view of the information contained in the news items and editorials published in *Gourdoot and Maldaha Samachar* during 1944 53 as well as in the '*memoir*' of Lalbihari Majumdar. In this our choice would be strictly confined to the information supplied by these local sources rather than any complete discussion of the themes.

Adivasi rituals and agrarian protest in Malda 1938-1947

The revolt of Jitu *Santal* in the *barind* region of Malda failed to attain its objective. [1] But for the first time, it drew the attention of the British authority to the untold sufferings of the *santals*. Shortly after the revolt, the District Magistrate proposed to government that a Special Officer should be appointed for the *barind* to enquire into the grievances of the *santals* and redress them as far as possible. [2] Sir P.C.Mitter, then the Revenue Minister, called for a report from the Settlement Department on the actual state of affairs in *barind*. On the basis of this report a Special Officer for *barind* was appointed in June 1933. [3] But this step did not usher any real reform. The *zamindari* oppression continued unabated. An enquiry in 1945-46 established that the *santals* in the Habibpur-Bamongola-Gajol complex of villages were remained sharecroppers with average holdings well below the size of an economic holding, and that even in the best of times their livelihood rose barely above the minimum subsistence level. [4]

So the conditions were ripe for further tenant unrest. In April 1938 violent clashes between the police and a tribal mob took place at *Darail beel*. [5] It was a large natural lake within the jurisdiction of Habibpur P.S. The *zamindar* of

Singhabad owned the *beel* which was known for the great quantity of fish it yielded every year. A large number of *santals* lived in the villages in the vicinity of *Darail beel*. On 6 April 1938 the *santal* tenants gathered around the *beel* to catch fish. The *Gourdoot* informed that this type of fish-looting was customary in the *santal* tradition.[6] They called it *bahich* and it usually took place after the *Fagua*.[7] The *santals* considered it their birth-right. Mr. Vas, then the Divisional Commissioner, ordered in the 1920s that the *santals* could observe this traditional custom at each pond for one day at *barind* region. Since then the *santals* had been observing this custom without any resistance from the *zamindars*.[8]

The *zamindar* of Singhabad sought the assistance of the police to prevent the *santals* from catching fish in Darail *beel*. A police force rushed to the spot and asked the *santals* not to catch fish in the *beel*. But the *santals* refused to obey. The officer-in-charge of the Habibpur P.S. fired a shot. The crowd became violent and attacked the police force and the *Peadas* of the lessee of the *beel*. A *Peada* was injured and later succumbed to death. Many polices were injured.[9] As the news of violence reached Englishbazar, the headquarters of the district, a large police force led by the Superintendent of Police rushed to the spot. The adjacent tribal villages were searched. The police arrested some leaders of the *santals*. A criminal case was filed against them. In his verdict, the assistant session judge of Malda sentenced four *santals* to rigorous imprisonment of two years each.[10]

Another conflict between the *santal* agriculturists and the local *zamindars* led to police firing on the unarmed *santals* at Chorol *beel* (pond) in April, 1947. It is interesting to note that all the standard literature on the Tebhaga Movement are prone to discuss this incident in connection with the Tebhaga Movement which broke out in 1946 in several districts of Bengal including Malda.[11] Our study however has shown that the Chorol *beel* incident had no relation with the Tebhaga episode. It was mainly a fight to assert customary rights of the *santal* community which the local *zamindar* tried to subvert.

From contemporary local newspaper source we come to know that while a large group of *santals* numbering six to seven hundreds were engaged in fishing with the help of *Polo* or *Tapa* (Pail-shaped wicker-basket for catching fish) at Chorol *beel* in the *zamindari* of Bulbulchandi, the police at the order of the *Barind* Officer [12] opened fire on them. It put, according to official estimate, four *santals* to death. [13] But the newspaper was unwilling to accept the official declaration of death-toll. The editor believed that the actual number of casualties was greater than the official estimate. The editor informed that he visited the place of event in person, made a thorough enquiry, interviewed the injured *santals* and some local people. On the circumstantial evidence he came to the conclusion that the official estimate of casualties was a deliberate lie. [14]

In support of his view the editor of *Gourdoot* furnished the following reasons : [15]

a) the police officials admitted that they fired thirty rounds of bullets. So it was unbelievable that only four *santals* were killed and a handful injured in view of the fact that the *santals* were unarmed and they gathered thickly at the pond. It left little chance for the police to be misaimed;
b) The police showed no intention to disperse the crowd by firing in air. They began shooting without any provocation from the *santals*;
c) The police had no intention either to hurt the *santals* by shooting at the lower part of body. The fact that two *santals* were found dead with bullet injury on their forehead confirmed this view;

d) Immediately after the *Adina* incident of 1932 (revolt of Jitu *Santal*) then the District Magistrate issued a statement on the event. But in the case of Chorol *beel* the district administration failed to issue any statement within two weeks after the incident took place. The newspaper considered this silence deliberate. It reported that the silence of the administration aroused suspicion in a section of local people.[16]

The newspaper reported that the local Communist Party, the District Congress and the *Hindu Mahasabha* unanimously condemned this brutal act on the part of the police in unequivocal terms.[17] The communists organized a procession in protest against police-firing on the unarmed *santals*. They demanded punishment of the police officials who were responsible for such brutality.[18] The leaders of the *Hindu Mahasabha* stated that this unfortunate incident could easily be avoided. They asked the administration to state clearly the situation in which the police were compelled to open fire.[19] The newspaper (*Gourdoot*) reported that conceding to public pressure the district administration had to issue a statement after a long spell of silence.[20]

The *santals* were, according to this statement, repeatedly warned by the *naib* of the Bulbulchandi Estate to stop fishing at the pond. While the armed *santals* paid no attention to his warning, he informed the police. A group of police and the *Barind Officer* rushed to the spot and ordered the *santals* to leave the place. But the *santal* mob refused to obey and a group armed with bows, arrows and spears attacked the police. Finding no other alternative the police had to fire in self-defense.[21]

But all the political parties, according to the newspaper report, expressed doubt regarding the authenticity of the official statement.[22] The local congress sent an enquiry team led by Ramhari Roy, M.L.A. to the spot. After a thorough investigation the enquiry team issued a statement to the press. The enquiry team was, the statement ran, in a position to say that the number of dead persons were fifteen to twenty. The police firing injured about forty to fifty people also. The mob consisted of *santals*, few *Rajbansis* and very few Muslims. The enquiry team stated that the police opened fire upon a peaceful gathering who were neither riotous nor they were armed with bows and arrows as alleged by the administration. Neither the mob threatened the police party with any attack. At the end of the statement the enquiry team vehemently criticized the role of the Barind Officer.[23]

The local Communists accused that the *Barind Officer* acted in collaboration with the *zamindar* and demanded his immediate resignation.[24] This demand was echoed by the newspaper also.[25] This brutal act on the part of the administration greatly stirred the mind of the local populace. This is evident from the reports of the local newspaper. The *Gourdoot,* for instance, had raised some questions[26]:

a) What was the nature of bullet injury, especially where it struck the body?
b) How and where the *santals* concealed their weapons while catching fish?
c) Did the *santals* escape after collecting their weapons or abandon those when firing began?
d) How many weapons were seized by the police from the spot?
e) During the *Adina* incident of 1932 many prominent personalities were employed by the administration to persuade Jitu to surrender. Did the administration adopt such type of persuasion before taking the ultimate path of firing in the case of Chorol *beel*?

The local intelligentsia was of the opinion that this brutal act of the police, aided by the local *zamindar*, was not an isolated event. They wanted to relate it to the deep-rooted agrarian problem of the *barind* region. In the editorial of *Gourdoot* it was stated that the incident of Chorol *beel* took place as the culmination of a long-standing conflict between the *santal* agriculturists and the local *zamindars*. [27]The principal causes of ill-feeling that existed between the landlords and the *santal* tenants of *barind* were the exaction of various *abwabs*, the illegal enhancement of rent, non-grantal of rent-receipts, the levy of interest at high rates and the rapacity and oppression of landlords' agents, *gomastas* and *peadas*.

This strained relation was particularly noticeable in the Bulbulchandi Estate. According to the *District Census Report of 1951*, the Bulbulchandi Estate was perhaps the most remarkable example of mismanagement in the district. [28] If the tenants refused to pay the enhanced rents claimed by the *zamindar* of Bulbulchandi, the unpaid enhancement had been entered as arrears in the estate accounts for many years and interest had been charged on it. Consequently it was not uncommon to find a *santal* tenant with an annual rental of Rs. 5/- had been shown in the estate accounts as being in arrears to the tune of Rs. 2000/. [29] The Bulbulchandi Estate also followed another malpractice, namely the deliberate destruction of rent receipts.[30]

These malpractices produced a bitter relation between the landlord and *santal* tenants in the *zamindari* of Bulbulchandi. [31] After the appointment in 1932 of a Special Officer at *barind* to look after the interests of the *santals*, it was hoped that this oppression would at least be reduced, if not ended. [32] But this hope of the *santal* tenants to get justice was dashed to the ground. It was seen that this Special Officer often cooperated with the *zamindars*. [33]

In December 1946 i.e., just before the incident of Chorol *beel* the *zamindar* of Bulbulchandi once again made an attempt to increase rents. [34] The *santal* sharecroppers protested vigorously and decided not to pay the enhanced rent. They also made an appeal to the *Barind Officer* to look into the matter. Several clashes took place and the *zamindar's gomastas* and *peadas* were assaulted on a few occasions. [35] The situation became tense and the *zamindar* of Bulbulchandi resolved to take revenge. The result was the incident of Chorol *beel* in 1947 where the *zamindar* with the help of the police gunned down some *santal* tenants on a trivial pretext. [36]

The occurrence of Charol *beel* is a rare incident in the panorama of general agrarian protests in Bengal. It shows how a ritualistic cause of tribal society could ignite a general discontent simmering within the framework of *zamindar*-peasant relations where rack-renting was the rule of the day. It also shows the complex relationship between different peasant groups, the tribal and the non tribal and how they were cemented when the onslaughts came. The incident of Chorol *beel* and its dating thus misguided some scholars who described a tribal community's bid to retain its customary right against the subversive attempt of the *zamindar* as a peasant struggle within the blazing trail of the *Tebhaga Movement*.

The strike of the malos in 1945

One of the most interesting information collected from the *Gourdoot* is perhaps the strike by the *malos* (fishermen) in protest against the fixation of prices of fishes by the district administration. In the context of the history from below, [37] this protest movement by the poor *malos* of Malda deserves attention and requires further serious research. The gravity and magnitude of the movement is revealed by the fact that the newspaper covered the news of this protest movement in four issues - 24 May 1945, 7 June 1945, 8 July 1945 and 6 September 1945.

The *Gourdoot* reported that the District Magistrate had fixed the prices of various kinds of fishes in the area under the Englishbazar Municipality. [38] The District Magistrate Mr. Barnwell further declared that this had been made according to the Safety of India Act. The District Magistrate warned the *malos* that in case of violation of the Act, strong punitive measures would be taken. [39] This step was editorially supported in the *Gourdoot*. The editorial also proposed that the fixation of rates be implemented in the entire district. At the same time some problems of the *malos* were pointed out - (1) rapid increase in auction value of ponds; (2) non-availability of fishes owing to drought; (3) enormous price-hike of essential commodities. [40]

Anyway, we come to know from the subsequent issues that some serious problems arose as a result of fixation of prices of fishes, such as -

a) The act was practically rendered void by the *malos*.
b) The prices of fishes went on increasing.
c) The *malos* were defying the law openly. They were selling fishes at a high price at their house and they were going outside the municipality area to sell fish at high prices.
d) As a result the influx of fishes at the *hats* and *bazaars* of municipality which was under the direct jurisdiction of the Act, was almost stopped. [41]

To solve this problem the district administration had declared that thenceforth the act would come in force in the entire district. [42] The administration had taken stern attitude and about 15 *malos* were arrested for violating the act. In the protest against this governmental action the *malos* called for a strike and the selling of fishes was stopped in the whole district. [43] The editorial of *Gourdoot* informed that the ordinary buyers were happy with the action taken by the district administration. They praised the administration for punishing the obstinate *malos*. [44]

The editor recapitulated that about 25 years back i.e. in 1920 a similar situation arose. [45] At that time a section of the congress workers led by one Ramlal Kabiraj called a meeting of the *malos* and ordinary buyers and fixed the price of different kind of fishes. The *malos* opposed this step in unison and a difference of opinion cropped up in the congress rank. Eventually C.R.Das, the state congress leader, had to be informed about this problem. He immediately expressed his concern and came at Malda only a few days after. He called for a meeting and in front of him the two sections of the congress who were for and against the fixation of prices of fishes began to make quarrel. [46] C.R.Das, after considering the opinion of the both sections, declared that the fixation of prices of fishes should not continue. In support of his view he cited several reasons :[47]

i) Uncertainty of catching fish. Though a *malo* caught three *maund* today, he might not be able to catch half *maund* tomorrow;

ii) Flexibility of demands. In marriage and other such occasions the demand usually went high. But at times the *malos* could not get buyers even at a low price.

The editor of *Gourdoot* was of the opinion that the problems of the *malos* cited by C.R.Das about 25 years back still continued. Moreover some other problems were added to worsen the condition of the *malos*, such as : [48]

a) The price of fishing boat was rocketed up after the War. Some three or four years ago it cost Rs. 30 - 32/- only. Its price now (i.e. 1945) stood at Rs. 100 - 125/-.

b) The price of fishing net increased by four times.
c) The auction value of ponds increased rapidly. The pond which some ten years ago was sold at Rs. 2,000/-, in 1945 its price now rocketed to Rs. 10,000 - 15,000/-

The editor stated that all these are the economic impact of the Second World War. So he requested the government to consider the present condition of the *malos* before fixing the rates arbitrarily.[49]

So we notice an interesting change of the stand in this issue. While in the issue of 24 May 1945 the editor enthusiastically supported the move of the local administration, now in September 1945 he came out in open support of the *malos*. We don't come to know why the newspaper changed its stand so quickly. We don't get any hint in the editorial either. Anyway, the newspaper opined that in case of fixation of rates, the interest of the poor *malos* should be duly protected.[50] The subsequent issues of the *Gourdoot* informed that the strike had been continuing. But rather surprisingly the successive issues did not contain any further information about the protest movement of the *malos*. It may be assumed that for being vocal in support of the *malos* the newspaper was warned by the authorities and it was refrained from containing any further news and taking up the cause of the *malos*.

Notes :

1. For a detailed discussion on Jitu Santal's movement in Maldah, see Tanika Sarkar, '*Jitu Santal's Movement in Malda, 1924-1932: A Study in Tribal Protest*', in Ranjit Guha ed., *Subaltern Studies IV*, New Delhi, 1985; Ashim Kumar Sarkar, 'Gourdooter Darpane Jitu

Santal o Barinder Santal Bidroho' (in Bengali) in *Itihas Anusandhan 9*, K.P. Bagchi & Co, Calcutta, 1997.

2. M.O. Carter, *Final Report on the Survey and Settlement Operations in the District of Malda 1928-35*, B.G. Press, Calcutta, p. 82.
3. *Ibid.*, p. 85.
4. K.P.Chattopadhyay, *Report on Santals in Northern and Western Bengal, 1945-46*, Calcutta, 1947, pp. 6-8.
5. Lalbihari Majumdar, '*Maldah Abalambane Jibansmiriti*', *Gourdoot*, 13 August 1949; M. S. Bhattacharya, *Studies in Micro history : Political Movement in Some Parts of India and Bangladesh, 1857 - 1947*, Indian Institute of Oriental Studies and Research, Kolkata, 2007, p. 85.
6. *The Gourdoot*, 1 May 1947.
7. One of the most important Santal festivities is *Fagua*. The *Fagua* is counterpart of the *Doljatra* festival. On the appointed day the villagers go to the place of worship and offer sacrifices. On the way back to the village they sprinkle each other with water, but it is considered highly objectionable to use colored water : see A. Mitra (ed.) *Malda : Census 1951, op. cit.*, p xxii.
8. *The Gourdoot*, 1 May 1947.
9. Case No. G 238 of 1938. Emperor Vs. Mangla Santal and others, file A 246, Malda District Records.
10. *Ibid.*
11. *See* Md. Abdullah Rasul, *Krishak Sabhar Itihas*, Nabajatak Prakashan, Calcutta, 1980; Siddhartha Guha Roy, *Malda*, Subarnarekha Calcutta, 1991; Sunil Sen, *Agrarian Struggle in Bengal, 1946-47*, Peoples Publishing House, Delhi, 1972; Suprakash Roy, *Bharater Krishak Bidroha O Ganatantrik Sangram*, vol. II, Calcutta 1969 91.
12. The Special Officer appointed after the revolt of Jitu Santal to enquire into the grievances of the *santals* was called the 'Barind Officer' by local people : see *Gourdoot*; 1 May 1947.
13. *The Gourdoot*, 17 April, 1947.
14. *Ibid.*
15. *Ibid.*
16. *Ibid.*
17. *Ibid.*
18. *The Gourdoot*, 24 April 1947; Narendra Nath Chakrabarty, '*Swadhinata Sangrame Maldaher Bhumika*', *in Radhagobinda Ghosh (ed.), Swadinata Sangrame Maldaher Abadan, Maldah, 1398 B.S.*, p. 7.

19. *The Gourdoot,* 17 April 1947; Sibendu Sekhar Roy, '*Swadhinata Sangrame Maldaha*', *in Radhagobinda Ghosh (ed.), Swadinata Sangrame Maldaher Abadan, Maldah, 1398 B.S.*, p. 9.
20. *The Gourdoot,* 24 April 1947.
21. *Ibid.*
22. *Ibid.*
23. *The Gourdoot,* 1 May 1947.
24. Narendra Nath Chakrabarty, '*Swadhinata Sangrame Maldaher Bhumika*', *op. cit.,* p. 11.
25. *The Gourdoot,* 1 May 1947.
26. *Ibid.*
27. *Ibid.*
28. *The Gourdoot,* 15 May 1947.
29. M.O.Carter, *Final Report, op. cit.,* p. 81.
30. *Ibid.*
31. *The Gourdoot,* 1 May 1947.
32. *Ibid.*
33. Based on interview with Manik Jha, then the secretary of the District Communist Party, 10 February, 1992.
34. *The Gourdoot,* 1 May 1947.
35. *Ibid.*
36. *Ibid.*
37. For a concept of '*history from below*' see Sabyasachi Bhattacharya, *History From Below,* in *Social Scientist,* April 1983, pp. 3-20.
38. *The Gourdoot,* 24 May 1945.
39. *Ibid.*
40. *The Gourdoot,* 8 July 1945.
41. *The Gourdoot,* 6 September 1945.
42. *Maldaha Samachar,* 8 October 1945.
43. *Maldaha Samachar,* 17 October 1945.
44. *The Gourdoot, 6 September 1945.*
45. *Ibid.*
46. *Ibid.*
47. *Ibid.*
48. *Ibid.*
49. *Ibid.*
50. *Ibid.*

Peasant and Artisan Resistance Movements of Bengal (1763-1300) : Revisited

Kartik Ch. Saha and Tapas K. Roy Choudhury

The whole range of views on the *Sannyasi* and *Fakir* Uprisings by Glazier[1], Firminger[2], Bankim[3], Ghosh[4], Ray[5], Chanda[6], Ray[7], Chatterjee[8], Dasgupta[9], Lorenzen[10] and Cohn[11], etc., make very interesting reading. Taken together they have explored the taxonornic position of the uprising from specific ideological perceptivity and therefore through paradigmatic shifts. Besides, in their quest for the roots to which the uprising could be attributed their arguments tended to have been underpinned by administrative, national and class interests. In their different constructions fusion of different categories of thwarted interests has emerged on very valid grounds Nevertheless, a closer scrutiny of the relevant primary sources suggests that a convergence between the different territoriality of the uprisings and the specific causes has determined the course of the movement with many finer shades. Despite Dasgupta[12] having explored the areas where the *Sannyasi/ Fakir* financial interests had been severely threatened by the exploitative policies of the colonial government at least two points, which we shall subsequently discuss, remained outside the consideration of the researchers. It has not been clearly defined either ii all those who participated in different operations, as mentioned in different official records with figures and broadly classified as *sannyasis* and *fakirs,* were the same although, or it was an ensemble of mendicants and secular civilians participating at different stages in the protest movements. Brides, a couple of important documents that we have come across in the U.P. archives while searching for documents relating to the siege of the Lucknow Residency, classified under Pre-Mutiny Papers (Banaras Correspondence) both at Allahabad and in Lucknow has urged upon us the necessity to qualify with them the contemporary debate. Further, our intensive scrutiny of the archival sources in the course of our study of the silk industry of Maid a and Murshidabad from prosperity to decline has enabled us to effectively argue that the revolutionary crowd was a complex structure of diverse interest groups which were uprooted by the colonial policies from the economic moorings at different points of time and by the rapid changes in the colonial encounter. They had reacted under the leadership of the fakirs and sannyasis for the organizational cohesiveness of the mendicants to negotiate a momentus phase of India's experience.

I

It was not an uncommon incident at all that the fakirs or the sannyasis sometimes lent, their services as mercenaries. Mir Kasim sought their assistance,[13] theywereusee by Khagendranarayan of Cooch Behar,[14] Darpa Dev of Jalpaiguri used them twice, once against the Cooch Behar ruler and later against the Collector of Rangpur when a dispute arose between them over the revenue assessment of his state.[15] And piracy, decoity perpetrated by some Bengal landlords through their retainers was also a regular-feature of the social life of the contemporary rural Bengal.[16] Glazier, strangely enough, made a category confusion by juxtaposing the incidents related to purely mercenary intent upon those whose primary objective was that of food riot, debt remission, non-payment of rent and retrieval of loan with accumulated interest. The uncertainty prevailed in all the dispatches of the military officials to their civilian counterparts in the affected districts namely Malda, Dinajpore, Rajshahi, Pabna, Bogra, Rangpore and Dhaka. We have no evidence of any fakir-sannyasi uprising either within the Cooch Behar State or in the Dooars partly controlled by Cooch Behar and partly by the Dev Raj of Bhutan.[18] On the contrary, the Eastern Morung, i.e., the territory between the rivers Mechi in the west and Mahananda in the east was seriously disturbed by the invading army of a subordinate Nepali military official namely Gangaram Thapa.[19] Just as the Dhimal jotedars namely Jitpal Maulik was adversely affected on account of this aggression, so was the neighbouring Baikunthapur estate of Darpadeva. Though it is mentioned in the Rongpore District Records that the Maharaja of Sikkim had gifted out five villages to Musa Shah, the rebel leader, in his morung territory the Sikkim records do not corroborate it.[20] Glazier did not ascertain the identity of Devi Chaudhurani, though he alone did mention her.[21] The question that always bothers is who was she? Besides, Devi's association With any landed interest except her surname which was used as a point of inference for the purported association was more interpolative in nature than factual.

Bankim Chandra eulogised her which is remembered by the nationalists, though his caution was either not known, or not seriously considered at all. Bankim candidly admitted in the advertisement of Armando. Math in his Banga Darshan that except the four names namely Bhawani Pathak,Devi chaudhurani, Goodland, and Lt. Brenan the remaining part of the story was a figment of his imagination.[22] Perhaps Bankim did not read anything mere than what was lifted by Hunter from Glazier's Report in his Statistical Account of Bengal Vol. 'IX and with which Bankim was associated as a contributor. The question is posed by us on account of the fact that another lady namely Jaydurga Chaudhurani whose name was more frequently referred in the Rangpore and Dinajpur/ District Records and who had already had a place in the Rangpore Rajbansi folklore[23] was not noticed by Bankim. Jaydurga was a landlord and was affected, among others, by the activities of the rentier Devi Singh who came as a revenue farmer with the Five-year Settlement of Warren Hastings. Jaydurga was a full-fledged rebel, but that she did not qualify for the 'Bandit Queen' imagery which could be attributed to Devi Chaudhurani is indeed mysterious.

The wealth of information that could be secured now on the episode under review was engendered in the course of researches undertaken in the last three decades or so. Jamini Mohan Ghosh certainly had long preceeded them, but emphasis on him could be laid more on the broad construction of the entire story than on detailed analysis of any specific theoretical issue. Besides, the tenor of his argument was overwhelmingly influenced by the British colonial historiography. The real classification of the revolt as 'peasant uprising',[24] 'Sannyasi, Fakir and peasants' war',[25] Peasant War',[26] 'Fakir-Sannyasi uprising',[27] indicate a real quest for the identity of the rebels, though the differences are patently clear. It is at this stage a new search is needed to further observe the differences in the objectives of protest launched in indifferent areas. Of the affected territories, as mentioned earlier, Malda, Dinajpore Rajshahi, Pabna and Bogra were more severely affected by the Famine of 1771-72 than Rang pore, Dhaka. Rang pore, Dinajpore Rajshahi and

Bogra were more distressingly affected by the new farming system then Malda and Pabna. And, Dhaka, Malda and partly Dinajpore and Rangpore were more severely affected by the investment policy of the East India Company in Silk and Cotton trade then the other districts. So, it is likely that the targets of the rebel attack and the morphology of the rebel crowd would be different in the three separate phases, i.e., (i) pre-famine phase (upto1769), (ii) period of famine (1769-72), and the period of five-year Revenue-Settlement of Warren Hastings (1772-1777) and (iii) the period of Ten-Year Settelement to Permanent Settlement (1777-1793).

II

The record of events in the pre-famine period indicates that the rebel crowd primarily targeted the English factories and arungs situated in some important cotton textile production centres. In 1763 a band of Fakirs attacked the Dhaka Factory and captured it. The Factory being abandoned by Mr. Leyester, the then Chief of the Factory until it was recaptured by Capt. Grant the Factory was systematically plundered by the rebels.[28] In the same year the Factory of Rampur Boalia in the Rajshahi District was plundered and its Chief Mr. Bennet was imprisoned and eventually killed.[29] Dhaka District Gazetteer referred to a note of Warren Hastings in 1763 that 'some time ago a large body of Fakirs infesting the countiy about Backergunj surrounded Mr. Kelley, my agent and put him in danger of life...the Fakirs have since quitted the country', though I have doubts about of the authenticity of the source, as Warren Hastings was not officially empowered at that point of time to employ any agent.[30] The assassination of two English officials, namely Mr. Myrtle and Lt. Keith by the Sannyasis in the morung in 1769 urged upon the government to initiate military operation against the Sannyasis.[31] For all practical purposes, the rebels thoughtfully focused on the East India Company's commercial establishments.

It is presumable that the changes in the investment policy of the East India Company after the battle of Plassey in 1757 and the gradual imposition of a monopoly on Bengal textile market by employing gomasthas to exact compliance to the Company order adversely affected the trading interest of the Sannyasis. A few other points too have been discussed by the scholers.[32] However; in the absence of any direct evidence from Bengal except one[33] (mentioning Cashmeerians, Multanys, Pathans, Sheikhs, Sannyiasys) the involvement of the Sannyasis, far less nanga sannyasis in silk and textile trade of Bengal cannot be established without the U.P. archival record. The information collected from the archives in Lucknow and Allahabad furnishes ample direct evidence of sannyasis trade, though they were nerally non-Bengali Sannyasi traders. Two representation from the Sannyasis, one by the Sannyasi merchants from Bengal trading with Deccan pleading for exemption of duties on goods imposed by the British[34] and another from the Nanga Sannyasis to the Resident of Benaras dated 1773 complaining against the excess duty charged on the merchandise[35] suggest a discriminatory tariff policy introduced by the British against the Indian merchants. In 1784 a petition was submitted to the Resident against fees collected by the Darogah and Officers of the Custom Houses established for levying the duties on the trade of the Sannyasis and Iraqi merchants.[36] In an identical fashion the merchants residing in the Mirzapur region complained to W.W. Playdell, Assistant to the Regional Resident to be communicated to Francis Fowke, Resident of Beneras complaining against the excess duty charged on the commodities.[37]

Again Birohi Dass and others, merchants at Mirzapur sent a representation in 1788 to the Resident of Beneras pleading for no imposition of levy on the goods brought from the south.[38] Further, in another communication Birohi Dass and other merchants pleaded to the Resident at Beneras in 1788 that inspite of their paying 5% regular duties they were subject to Khanashumari duty.[39] Another report from the Magistrate of Mirzapur showed that one Kallendar Gir paid duties on stores according to government regulation.

From the new archival information furnished above a number of significant points emerge. First, it is realised that the British policy to impose levy on the Indian merchants in the matter of transhipment of goods from one region to the other was considered by the Indian merchants as both differential and therefore detrimental to their business interest That the British refused to respond to their request presumably added a strong *around* for them, along with others, to try to subvert the colonial imposition. It is further known that the merchants might have defied payment of the levy for some time and the resistance might have collapsed in 1790. Payment of 5% regular duty,as mentioned in one document, is an important piece of information that suggests that the British contrived to inflict mortal injuries to the Indian traders after the Chait Singh affair. The involvement of the sannyasis in Bengal's textile and silk trade could be directly established now, however, Puran Giri Gossain's name is often mentioned to substantiate the point particularly on account of the fact that he was asked by Warren Hastings to accompany Bogle as an interpreter for his negotiation with the Tashi Lama.[41] Further, the Gosain traders of Beneras mentioned by Misra and also referred to by Cohn did not remove the uncertainty. However, what is interesting is that fakir association with any trading is not found to satisfactory explain why did they, as mentioned earlier, actively participate in the raid of Dhaka and Rampur Boalia factories. (The interference in the collection of contributions to meet the expenses of pilgrimage came only after British acceptance of the Dewani in 1765, the resumption of tax free madad-i-rnasha, Nazoorat, Nazir-i-Dargah, Brahmottar, Debottar, Sibottar etc., originated from 1788 regulation and therefore, could not be having anything to do with 1763 raids.) Hence, it may be presumed that the use of the term 'fakir' in the official correspondence with reference to the 1763 incidents was in fact category confusion. The madari Sufis, i.e., the besarias in their appearance corresponded to those of the sannyasis, and sometimes eclectic names were used such as Kalandar Gir obtained from one U.P. document explain the confusion of the British.

> From 1771 to 1790 report of twenty four uprisings could be secured from the District records and other categories of archival materials, of them, fifteen uprisings took place in Rangpore, Bogra, Dinajpore, Rajshahi and Mymensingh. In 1771 the first skirmish between Majnu Shah, the fakir leader[42] and Lt. Feltham was reported from Bogra.[43] Almost simultaneously, another incident from Erdakpore of Rangpore was reported.[44] It was also reported that land-holders of Silberiain Rangpore helped the rebels.[45] In 1771 sannyasis were reported to have entered into the Alapsing subdivision of the district of Mymensingh.[46] In 1772 a party of fakirs was reported to have plundered the kutchery (rent collection office) of Jaisindhu subdivision of Rajshahi district plundering Rs. 1000.00.[47] A joint operation of Sannyasis and fakirs in collusion with the zamindars was reported in 1773.[48] The plunder of the kutchery of the zamind'ar of Mecheda in the Dinajpore district was reported in 1773.[49] Disturbances in revenue collection was brought to

the notice of the Court of Directors in 1774.[50] Interestingly enough, incidents of the recruitments of the Rajputs and the locals by Manju Shah was noticed by the British officials in 1776.[51] The news that the rebels did confine the naibs (revenue official) and kidnap zamindars also came to light from the Collectors of Dinajpore and Bogra in 1773.[52] Further incidents of confinement of the naibs and gomastas of the zamindars in the District of Mymensingh were reported in 1773.[53] Often it was admitted by government functionaries that '...the depredations they have committed in Silberis etc. to have been chiefly owing to the infatuation of the Talookdars, Suhabitans etc.in not giving proper intelligence in countenancing 'and abetting these plunders'.[54]

Incidents of plunder and confinement took place even after the Five-year Settlement was revoked and later it was being replaced by the Ten-year Settlement of Cornwallis. The Gomatas of the zamindars of Alepsing in the Mymensingh district were plundered by Majnu Shah in 1780.[55] Majnu Shah had also kidnapped the son of the zamindar.[56] It was at this stage that the uprising spread in the district of Malda. Attack on the kutchery of Pakura in the Malda district by Majnu Shah was reported by the Chief of the Malda factory.[57] He also reported the attack on the Nirschindapore factory of the English.[58] Report of the plunder of revenue by the fakir at Mymensingh was reported in 1786.[59] Almost at the same time report was received that the Tehsildar of Gilabary in the Dinajpore district was kidnapped.[60] Again revenue from Jaffarshahy of the Mymensingh district was plundered in 1790 by Cherag Ali, one of the successors of Majnu Shah.[61] Attack on the Company factories did not cease; even in 1792 Birtara factory of the Company in the Mymensingh district was plundered.[62] The revolt continued even after the introduction of the Permanent Settlement in 1793. Because, an incident of plunder having taken place at Nawabganj in 1794[63] suggests that the rebels certainly were not satisfied with the new revenue policy. Or, the complete estrangement from their traditional sources of subsistence induced them to sustain their armed protests till they were totally destroyed. From the preceding study it tends to appear that a good many armed protests took place during the period' of the Five Year Settlement, i.e., from 1772 to 1777. And the areas which were seriously disturbed by the new farmers like Devi Singh came up for the targets of the rebels. The recurrence of uprisings in the Districts of Rangpore, Mymensingh, Dinajpore, Bogra *etc.* largely explain that the traditional zamindars and the peasants were hit hard by the policy of inconsiderate exaction of rent by the new revenue farmers. Aggression on the Company factories during this period being less than that of the previous decade was not fortuitous at all. On the contrary, the entire range of operations was tailored to the interest of the zamindars and the peasants more than it being focussed on the interest of the trading communities, the weavers, the moneylenders and the cotton and mulberry cultivators. The decline of the cotton textile industry had already set in, and the silk industry was gasping for breath due to sudden change in emphasis by the Company from textiles to raw silk. Besides, the discontent of the weavers may be attributed to certain policy decisions which the officials of the Board of Trade initiated to effectively control the weavers. The situation cropped up on account of the decline in the textile price caused by the high-handedness of the *gomastas* and sometimes their contrivances being endowed with the support of influential government functionaries like Barwell.[64] 1757 *batta* and *dalali* to the tune of 3.3 per cent was deducted from the total price paid to the weavers as a regular mode of transaction. Bolts noted it in his *Considerations.* According to Bebb, the Dhaka Factory Chief after Hurst, and who was referred to by Sinha[65] the price of every commodity needed for household management and for manufacture went up by 20% to 30% with reference to the price index of 1757 and the textile price offered was less than the price index of the base year of 1757 in an inverse proportion. Certain freedom were accorded to the weavers in 1773 in consideration of the decision of the Court of Directors in 1771 towards granting the liberty to sell textile products in an open market.[66] However, temporary liberality' shown to the weavers was later withdrawn and some draconian measures were imposed on them to restrain them from going to the open market after entering into any kind of contract with the East India Company, or having taken *dadan* (advance) on that account. Regulations preventing the weavers to go to the open market were imposed on 22nd April, 1782, 19th July, 1786, 23rd July, 1787 and the supplementary to it passed on 30th September, 1789. On individual effort Hurst arranged for *some relief to the Dhaka manufacturers in 1778-1779,67 but it could not be sustained on account of the broad policy changes mentioned above. Presumably, a large number of weavers who were severely affected by the exploitative policy of the colonial administration were likely to be totally alienated.*

The condition in the silk industry sector turns out to be different from that of the cotton textile, as discussed above. After the 1770 famine and the complete dislocation of the industry some concrete measures were taken in pursuance of the recommendations of the Committee of Circuit[68] *to gainfully rehabilitate the tutias or tut chashis. Rent free land grant for two years and for the third year a relief to the tune of fifty per cent of the pre-famine rent per bigha enabled the mulberry planters to step up production and a remarkable progress was noticed by 1779. But the grounds gained could not be sustained due to the European wars in 1780-83 and later due to fall of the silk price. Sinha referred to the loss in absolute terms to the extent of Rs. 4.10 per bigha suffered by the tutias.*[69] *Almost simultaneously with the reverses suffered by the tutias the nakauds, chashars ttcj (the reelers) experienced a change in fortune for some policy changes from 1774 to 1783. Till 1773 a remarkable difference in price between Bengal raw silk and filature silk after filaturing was introduced by Weiss in 1771 had continued to exist. But from 1774 a demand for raw silk had suddenly escalated resulting in price adjustment between Bengal raw silk and filature silk between 1765 to 1773 and it was recorded as Rs. 6.80 And 10.16 Respectively. During the subsequent one decade, i.e., from 1774 to 1783 the price of Bengal raw silk varied from Rs. 9.16 to 7.75 per seer, while that of filature silk declining from Rs. 14.00 to 8.75 per seer.*[70] *Besides the European market, Bengal raw silk market in Northern India and in the Deccan remained virtually unaltered. From the Mangee Customs House Record Sinha secured the information that the export of silk in North India in 1789 amounted to Rs. 19.96 lacs and in 1790 the export declined to Rs. 16.96 lacs.*[71] *The benefit accruing from this trade went primarily to the intermediaries than to the primary producers. Besides, private filatures came into existence and silk produced annually went to Mirzapore, Nagpur, Lahore, Multan. From one centre namely Boaleah one thousand maunds of raw silk was produced* amounting to Rs. 3.54 lacs. This information presumably does not correspond to the argument tendered by Suranjan Chatterjee relying on information furnished by K.M. Mohsin, G. Bhadra etc. It is true that in course of six decades for which we have made the

> calculation, i.e., from 1705 to 1767 the business of the native traders in raw silk and silk piece goods declined from 1037 tons to 223 tons, but some change towards the better was noticed in 1789, 1790 when raw silk was found to have been exported to North India on the basis of the native traders from one Customs House only. The total quantum of exports were 206.2 tons and 173.7 tons respectively. The export figures from other Customs Houses and to Deccan certainly could have given more comprehensive picture than the present one. Some indication of change can be noticed In the import figures of Banaras and Mirzapur between 1775-1777. Silk piece goods imported for Benaras were 3,520 pieces and 21,267 pieces were imported for Mirzapur. The imports of raw silk were 533 maunds (19.39 tons) and 11,961 maunds (434.95 tons) for Beneras and Mirzapur respectively. Cotton piece goods imported were 15,801 and 81,082 respectively for the twocities.[72]

The two acts of the colonial government, viz., the termination of the Tax-free land grants of the pre-colonial government and the interference in the collection of pilgrimage subscription presumably caused wide-spread dislocation in the economy of the mendicant class. A broad idea can be formed about the number of persons involved in the cultivation and management of the revenue free land on the basis of acreage of land classified as tax-free tenures in the pre-Dewani settlement period. Depending on the study of Kamalakanta Gupta[73] R.M. Eaton[74] calculated the total revenue-free tenures in Silhet from Aurangzeb (1658-1707) to Alamgir II (1754-1759) as 908 involving 50,566.05 acres. In the district of Bogra thetotal rent-free tenures was 16,444 taluks. This calculation was of course confusing on account of the fact that revenue-free pre-dewani settlement grants and the post-Permanent Settlement rent-free grants of the zamindars were not separately denominated. In the District of Dinajpore the pre-colonial grants accounted for 178 taluks and 33,904 acres. [76] All these grants were made by way of implementing reclamation of wasteland and conversion of them into arable land to enhance revenue income of the state. Religious persons were fruitfully engaged to draw labourers of specific nature to perform these act. [77] After the 1788 Act towards the cancellation of the revenue-free tenures understandably a severe reaction would take place and the colonial administration responded to it by a proclamation in 1788 itself that the pre-17.65 revenue-free tenures would not be resumed; nevertheless, neither of the decisions was given effect to, [78] Consequently, it was unlikely that those who subsisted from revenue free tenures, comprising the original mendicant grantees and the peasant under-tenure holders, were disturbed in their subsistence by the British resumption policy. The other cause of alienation of the fakirs and sannyasis, i.e., the interference of the government in the collection of pilgrimage contributions, and complained against by the British officials as forceful exactions needed detailed scrutiny. Till 1778 there is virtually no record of any complaint from the. Natives to the British officials against fakirs/sannyasi pilgrimage subscription as 'exaction'. The letter of Majnu Shah alleged to have been sent to Rani Bhawani and quoted by Jamini Mohan Ghosh which understandably was used as evidence of-British interference, although there are others in the reports of different collectors, could not be traced by us. So, it is seen that the effect of governmental interference in the matter of religious practices on the ground that the religious congregations enabled the pilgrims to organise armed uprisings had very focussed effect than the economic dislocation caused by the policy of resumption of revenue free grants. The alliance between the Fakir-Sannyasis and the zamindars and peasants became patently clear from different sources.[79]

Ill

What transpires from the explorations of the facts related to the 18lh century uprisings is that the morphology of the revolutionary crowd was not uni-linear as it was made to be by some scholars. The composition of the crowd varied in terms of the economic dislocation suffered by different segments of the population and on account of different policies applied by the colonial government. From 1763 to 1772 the crowd appeared to be composed of the fakirs, sannyasisi and primarily the cotton weavers who were economically marginalised by the introduction of the gomastha system, imposition of levy on native traders and the arbitrary reduction of the textile prices. The incidents of attack on the English factories and araungs particularly in the textile districts testify to this assumption and the absence of incident of plunder of the - kutcheries of the zamindars, which was more frequent in the second phase of the uprisings further strengthened the argument. The period of Five Year Settlement, i.e., from 1772-1777 witnessed the most intense period of the uprising which spread over Rangpore, Mymensingh, Dinajpore, Bogra and Rajshahi. Presumably, the revolutionary crowd had a more complex composition than of the previous period comprising the sannyasis, fakirs, small zamindars, peasants etc. The incidents of revolutionary activities could be typified by acts of plunder, exaction of ransom money from the revenue officials, - from the landlords, disturbance in rent collection etc. During this phase of the movement no incident of factory attack did take place, though, during the subsequent period, when the Five-year settlement was withdrawn and new settlements until the Permanent Settlement were promulgated, the composition of the revolutionary crowd remained virtually the same except the zamindars. The reference to some disturbance at Malda in the British records can be attributed to the decline of private trade in silk. The fall in the price of mulberry leaf around 1783-84 caused serious discontent amongst the mulbeny cultivators namely tut chashis who might be forming the warp and woof of the revolutionary crowd that plundered the - kutchery at Pokua and the ourang at Nirschindapore at Malda in 1783. At this stage the weavers too might have again participated in the resistance movements in a state of desperation on account of the repressive regulations.

It is indeed mystifying that no incident of food riot had been reported by the Company officials during the period of 1770-72. Interestingly enough, in certain passages of Majnu Shaher Hakikat written by Jamiruddin Dafadar and referred to by Dasgupta[80] explicit reference to the question of food had been made. Majnu Shah was stated to have been explicitly implored by the khadim of the khanka to plunder grain to save the life of starving villagers. Nevertheless, incidents discussed above were related to textile and silk, production and trade and the revenue settlements of the East India Company. Official policy to store gram was taken in the post-famine period, but it is unlikely that stores of the local grain dealers were not invaded by the famine striken farmers. On the whole, the composition of the revolutionary' crowd was not monolithic, and the specificity of the crowd depended on the specificity of the economic dislocation in specific industries and in the land control system of Bengal in varied locale and social stratum of the society.

The theoretical issue that was bothering the scholars regarding the classification of the movements can be reasonably solved on empirical evidences. Ranjit Guha preferred to exclude the eighteenth century uprisings from the arena of subaltern movement on the ground of inadequacy of information.[81] The plea of inadequacy being no longer tenable it transpires that varied compositions of the revolutionary crowd, at different points of time with the sannyasis and Fakirs as the common element, tended to suggest the terminological multiplicities in classification. The peasant factor was no doubt the most important populist dimension of the movements since 1772; however, it is difficult to infer any peasant participation in the 1763 factory raids. Further, the Sannyasi and Fakir elements have posed a serious taxonomic problem on account of the fact that they did not conform to any class category. Except being intermediaries in the export trade and being involved in usury the identification of this category as primary producers is difficult to substantiate. Occasional incidents of pirana, naukarana, Devottar, Siuattor, Brahmattor property being cultivated on khudkhast basis do not qualify the Sannyasis and Fakirs to be categorised as producing community. Usury is being always viewed as an act of exploitation and on this account the Bengal Left could not lend its support to those who opposed the Debt Settlement Act of the Fazlul Haque Government in Bengal in 1937. Due to non-productive socio-economic roots of this community, their class character cannot be confidently identified as something commensurable to that of the artisans and the peasants. They of course played a role in the uprisings, identical to those of the Sans Culottes in the

French Revolution. Besides, the majority of the sannyasis and fakirs were non-resident, non-Bengali itinerant peddlers. So, the act of identifying the movements with their unrest would be a travesty of truth. The Fakirs and Sannyasis resembled some very significant external features of the French sans culottes on account of the fact that both lived in slums, khankas and maths were no better than slums, and did not wear any other garments under their shirts or chadars, such as the sans culottes did. In view of the problems discussed it would be too simplistic to describe the uprisings either as Fakir and Sannyasi uprising or as Peasant War or Sannyasi Rebellion. [82] To *classify the uprisings, therefore, as Peasant-Artisan resistance,* though *this taxonomic terminology too is somewhat imprecise, is more true to the contemporary construction of the events than otherwise. Hence, for all practical purposes, the movements ought to be attributed to the combined resistance of the artisans and the peasants of Bengal, whose suffering exceeded that of any other category of participants and who formed the bulk of the revolutionary crowd till the resistance collapsed.*

Notes :

1. E.G. Glazier, *A Report on the District of Rangpur*, Vol.' I, Calcutta," 1873.
2. W.K. Firniinger (ed.), *Bengal District Records 1763-1774*, Rangpur, Dinajpore.
3. Bankirn Chandra Chattopadhya, *Bankim Rachanabali* (ed.) Jogesh Chandra Bagal, Vol. 1, Calcutta, 1964.
4. Jamini Mohan Ghosh, *Sannyasis in Mymensingh*, Dacca, 1923, Sannyasi and Fakir Raiders in Bengal, Calcutta, 1930.
5. Suprakash Roy, *Bharater Krishak Bidroha O Ganatantrik Sangram* (in Bengali), Calcutta, 1966.

6. A.N. Chanda, *The Sannyasi Rebellion*, Calcutta, 1977.
7. N.B. Roy, '*Naga Sannyasi Ganeshgeer and the Kutch Bihar disturbances of 1787*' in *H.R. Gupta (ed.), Jadunath Sarkar Commemoration Volume*, Vol. 2, 1958, pp. 275-80.
8. Suranjan Chatterjee, '*New reflection on the Sannyasi, Fakir and Peasants' War*' *in* Economic and Political Weekly, Vol. XIX, No. 4, January 28, 1984.
9. Atis K. Dasgupta, *The Sannyasi and Fakir Uprisings*, Calcutta, 1992.
10. D.N. Lorenzen, *Warrior Ascetics in Indian History*', Journal of American Oriental *Soceity*, Vol. 98, No. 1, 1978, pp. 61-75.
11. Bernard S. Cohn, *The role of Gosains in the economy of the 18th and 19th century Upper India*', Indian Economic and Social History Review, 1, No. 4, 1964, pp. 175-182.
12. Dasgupta, op. at, Chapter 2, Unfolding Contradictions. The points mentioned by Dasgupta had been previously discussed, more or less on the same web-length, by Suranjan Chatterjee.
13. Ghulam Hussain Tabatabai in his *Siyar-ul-Mutakhkherin* (translated by Haji Mustafa), 4 vols., Calcutta, 1926, spoke about how the Sannyasi mercenaries were inducted by Mir Kasim to confront the army of the East India Company.
14. N.B. Roy, *op. cit.*, has discussed the role of Ganeshgeer to enable Khagendra-narayan of Cooch Behar to combat the Royal forces in the course of a conspiracy for the restoration of Khagendranarayan as Diwan.
15. Regarding the induction of the Sannyasis by Darpa Dev against the Nazir Deo of Cooch Behar the details can be secured from the Letter of the Collector of Rangpur to the Committee of Circuit, dated 12 January, 1773. The British action against the Sannyasi mercenaries of Darpa Dev had been outlined in the Letter from Capt. J. Jones to Warren Hastings, dated 27th, 28th January, 30th January, 1773. Article 8 of the 1773 treaty between India and Bhutan on Cooch Behar affair also may be mentioned.
16. Bankim Rachanabali, op. cit. Sir Jadunath Sarkar in one of his writings had dealt with the involvement of some zamindars in the acts of banditry.
17. Glazier, *op. cit.*, Classification of all the uprisings as 'banditri' originate from the refusal of the British officials in general to understand the intent of the unrest.
18. On a closer scrutiny of the incidents of Fakir and Sannyasi involvement in Cooch Behar during the time of Khagendranarayan or during the time when the King

Dharyendra Narayan was kept as a hostage by the Dev Judhur of Bhutan the mercenary character of the Gosains was more pronounced than any protest motivation which had induced them to fight against the British in other districts of North Bengal.

19. Tapas K. Roy Choudhury, 'Eastern Morung : A Disputed territory in Indo-Nepalese Relations', *Indian Historical Review*, Vol. xvii.
20. Land given to Musa by Sikkim is reported in the Letter from Capt. Alexander to the Collector of Rangpore, 25th March, 1786, Rangpur Records, Vol. 33, p. 45 and also referred to by A.N. Chanda. In the course of my editing the History of Sikkim by Maharaja Thutob Namgyal and Maharani Dolma, the only official history of Sikldm we have not come across any reference to this grant, nor did they mention the aggression of Gangaram Thapa on a part of the territory which theoretically belonged to Sikkim at that point of time.
21. Glazier alone mentioned Devi Chaudhurani and he too did not mention the name of her husband nor the territory which they used to hold as . the landlord.
22. For details see the advertisement of Ananda Math written by Bankim in his Banga Darshan, Baishakh, 1291 B.S. (1886).
23. Joydurga Chaudhurani, the wife of Shib Chandra, the Zamindar of Itakumar, situated at the border of Rangpur and Dinajpur rose in revolt against Devi Singh, the new revenue farmer and was mentioned both in the Rangpur and the Dinajpur Records. A local poet namely Ratiram Das eulogised her in his poetical composition, which was subsequently reprinted in the Rangpur Sahitya Parishad Patrika, 1315 B.S. (1900) and was used by Narahari Kaviraj in his A Peasant Uprising in Bengal 1783, Calcutta, 1972. Joydurga of course revolted during 1772-77.
24. Narahciri Kaviraj, *ibid.*
25. Suranjan Chatterjee, *op. cit.*
26. B. Chaudhury, The Transformation of Rural Protest in Eastern India, 1757-1930, Proceedings of the 40th Session of the Indian History Congress, 1879.
27. Atis Dasgupta, *op. cit.*
28. Secret Department Proceedings, dated 5.12.1763.
29. Letter from the Collector of Laskarpur to the President of the Council, dated 4.4.1773.
30. B.C. Allen, District Gazetteer of Dacca, 1912,pp. 40-42.
31. K.D. Bhargava (ed.), Fort William Indian House Correspondence, Vol. VI, Delhi, 1960. Public Letters dated 25 January, 1770, p. 177,

Proceedings of the Controlling Council of Revenue at Murshidabad, Vol. XI, pp. 35, 58.

32. Chanda, Chatterjee, Dasgupta, op. cit.
33. William Bolts, *Considerations on Indian Affairs*, London, 1772, p. 193.
34. Pre-Mutiny Papers (Beneras Correspondence) Appendix 13, Vol. 179. Information used by Conn and J.P. Misra vide Benaras in Transition,- deals with a later period. Our sources give information of 1773. Besides, C.H. Barlow submitted his Report in 1787.
35. Ibid. Appendix 7, V, 164-65; Appendix 6, V, 155-57.
36. op. cit. Appendix 13, v, 179.
37. Banaras Correspondence, v, 58-60.
38. Banaras Correspondence IX, 25. From Hasting's report to the Council we come to know that tire rates imposed by him to equalise the rates between the British and the Indian merchants were the following:

Commodity	Per maund		
	Rs.	Annas'	Piece
Silk and silk cloth	3	13	4
Cotton cloth (per bale)	1	12	2

Source: Bengal Secret Consultations, 1773, Range A, 21, p. 535.

39. *Ibid*, IX, 27.
40. *Ibid*, XXXill, 261-264.
41. C.R. Markham (ed.), *Narrative of the Mission of George Bogle*, London, 1876.
42. Regarding Majnu Shah the British official records referred to only those incidents which were related to the uprisings. The relatively complete story was drawn a litde over hundred years later by one Jamiruddin Dafadar from Birbhum. Majnu Shaher Hakikat therefore is interesting, though it is a very weak piece of historical evidence.
43. Letter of Capt. James Rennel to the Controlling Council of Revenue at Murshidabad, 1st March, 1771, Letter from the Supervisor of Rangpur dated 28th February and 7th March, 1771.
44. Letter from the Supervisor of Rangpur to the Controlling Council of Revenue at Murshidabad, dated 7th November 1771.
45. Proceedings of the Controlling Council of Revenue at Murshidabad, dated 19 and 27 April, Vol. III. Letter from the Supervisor of Rangpur to the Controlling Council of Revenue dated January, 1772.

46. Letter from the Supervisor of Rajshahi to the Controlling Council of Revenue at Murshidabad, 23 January 1772.
47. *Ibid.*
48. Secret Department Proceedings, dated 15 March, 1773.
49. Letter from the Collector of Rajshahi to the Governor General in Council, 8 December, 1773.
50. Extract from Revenue General Letter to the Court of Directors, dated 18 October, 1774.
51. Proceedings of the Provincial Council of Revenue at Dinajpur, dated 18 June, 1776.
52. Letter from the Committee of Circuit at Dinajpur to the Collector of Scolberis, Bogra, 13 January 1773, Secret Department Proceedings, dated S and 21 January, 1773.
53. Letter from the Collector of Dacca, dated 26 January 1773.
54. Original Consultation No. 9, dated 20 July, 1773.
55. Proceedings of the Provincial Council of Revenue at Purnea, dated 14 March, 1780.
56. *Ibid.* 1780.
57. Letter from Charles Grant, Resident at Malda Factory to Augustus Cleveland, Collector of Bhagalpur, dated 8 March, 1783. Revenue Department Consultation, No. 14, 9 March, 1783.
58. Public Department Consultation No. 13 and 15, 1983.
59. Committee of Revenue Proceedings, Original Consultation, Nos. 16 to 18, 1786.
60. Bengal District Records, Dinajpur Vol. II.
61. Revenue Department, Original Consultation No. 60, 1790.
62. Ibid. Original Consultation, No. 28, 1792.
63. Extract from the Judicial Department (Criminal), Original Consultation No. 11, 1794. Besides, information relating to attack on big landlords could be secured from Proceedings of the Governor General in Council, Political Department, 1795.
64. N.K. Sinha, *Economic History of Bengal*, Vol. I, Calcutta, 1965, pp. 159-60.
65. *Ibid*, p. 161.
66. Proceedings of the Board of Trade, 20 July, 1786.
67. Sinha, *op. cit.*, vol. I, p. 194.
68. From the Proceedings of the Board of Trade, N.K. Sinha calculated the loss suffered by the mulberry growers on a bigha of land. The market price of leaf plucked from one bigha of land was Rs. 9.75 while the total production cost and rent together came to around Rs. 13.85.

69. N.K. Sinha, Economic History of Bengal, Vol. I, p. 198.
70. Mangee Customs House Records, 1789.
71. N.K. Sinha, *op. cit.*
72. For details see the author's paper 'Indigenous Industry : India's Phenomenal success through the Silk Route', Journal of Historical Studies, No. 3, 1997.

Year	Indian Traders' Annual Average in tons	Year	Indian Traders' Annual Average in tons
17th c.	1037	1758	673
1748	1037	1759	533
1749	742	1760	484
1750	725	1761	393
1752	653	1763	225
1753	669	1764	309
1754	565	1765	267
1755	455	1766	191
1756	283	1767	223
1757	291		

Sources : Geogheghan, Some Account of Silk in India, Calcutta, 1872.

(ii) Milburn, Oriental Commerce, Vol. II. London, 1872.

(iii) Prog, of the Board of Trade, No. 27, Vol. 221.

(iv) Reports and Documents connected with the Proceedings of the East India Company in regard to Trade, Culture and Manufacture of Raw Silk, London, 1836.

(v) Calcutta Customs House Records, 1871.

Bengal Board of Revenue (Miscellaneous) Proceedings, Range 98, 18, 20 and 22 (1975-77) collected by J.P. Misra vide Banaras in Transition, p. 162.

73. Kamalakanta Gupta, *Srihater Bhumi O Rajasva Babastha*, Sylet, 1966.
74. R.M. Eaton, *The Rise of Islam and the Bengal Frontier*, Delhi, 1997, pp. 260-61.

    The total number of grants were spread over Brahmottar, Devottar, Vislinottar, Sivottar, Madad-i-Masha, Hindus, Muslims, Chiragis.
75. M.M. Choudhuri (ed.), *Bangladesh District Gazetteer* (Bogra), Dhaka, 1989, pp. 256-257.
76. F.W. Strong, *Dinajpur Gazetteer*, Allahabad, 1912, p. 130.
77. This argument has been tendered by Eaton vise The Rise of Islam and the Bengal Frontier. Besides, the episode dealing with the settlement of a city by Kalketu in Mukundaram Chakraborty's Chandi Mangal Kavya underpins the role of the charismatic

religious personalities like the Sufi saints in mobilizing the work force, the bosunias to reclaim the forest land.

78. N.K. Sinha, has summarised the reaction of different Collectors regarding Baze Zamin calculation and the likely impact that the implementation of the revenue free land resumption policy would provoke vide Economic History of Bengal, Vol. II, Appendix D. It transpired from the Prog, of the Board of Trade and from the communications of the Collectors that a confusion had prevailed in their differentiation between madad-i-mash, altamgha, Piran, Brahmottar, Debottar, Sivattor etc. Pre-Diwani revenue-free grants and post-Dewani lakhiraj grants. In all cases the pattas granted to the subordinate tenure holders were of the nature of mourasi-mukarari type involving one time lump-sum rent payment. In most tenurial allotments annual rent per bigha was fixed at Rs. 1.00 with exemption from the government cesses. Dr. Ratan Dasgupta studied the documents of the Bais Hazari Estate of the District of Malda and came to the conclusion that the lessees always resisted successfully the attempts of the grantees, i.e., mutwalis in this case to enhance rent on the grounds of land valuation in the adjoining areas, increase in crop production, increase of crop price and improvement in marketing facilities enabling the rayats to obtain better price. For some relevant information see J.J. Pemberton, *Report of the District of Maldah*, Calcutta, 1854, M.O. Carter, *Final Report of the Survey and Settlement Operations in the District of Malda*, Calcutta, 1938.

About the Contributors

Anjan Chakrabarti, Assistant Professor in Economics, St. Joseph's College (University Section), North Point, Darjeeling - 734104.

Dr. Sahara Ahmed, Assistant Professor, Department of History, Rabindra Bharati University, Kolkata.

Dr. W. Jyotirmoy Singh, Senior Lecturer, D.M. College of Teacher Education, Imphal, Manipur.

Dr. Kaushik Chakraborty, Lecturer, P.G. Department of History, Malda College.

Dr. Sujaya Sarkar, Associate Professor, Department of History, Vidyasagar University.

Ramendra Nath Bhowmick, Assistant Professor, Department of History, Shamuktala Sidhu Kanhu College, Coochbehar.

Dr. Sabyasachi Dasgupta, Assistant Professor of History, Visva-Bharati.

Manas Dutta, Assistant Professor, Department of History, Harishchandrapur College, Malda.

Dr. Pinaki Roy, Assistant Professor of English, Malda College.

Dr. Ashim Kumar Sarkar, Associate Professor of History and Teacher-in-charge, Malda College.

Kartik Ch. Saha, Assistant Professor of History, Bakshirhat College, Coochbehar.

Tapas K. Roy Choudhury, Former Professor, Department of History, North Bengal University.

INDEX